Modeling
and Control
of Precision
Actuators

Modeling
and Control
of Precision
Actuators

Tan Kok Kiong • Huang Sunan

CRC Press
Taylor & Francis Group
Boca Raton London New York

CRC Press is an imprint of the
Taylor & Francis Group, an **informa** business

CRC Press
Taylor & Francis Group
6000 Broken Sound Parkway NW, Suite 300
Boca Raton, FL 33487-2742

First issued in paperback 2017

© 2014 by Taylor & Francis Group, LLC
CRC Press is an imprint of Taylor & Francis Group, an Informa business

No claim to original U.S. Government works

ISBN-13: 978-1-4665-5644-7 (hbk)
ISBN-13: 978-1-138-07247-3 (pbk)

Visit the Taylor & Francis Web site at
http://www.taylorandfrancis.com

and the CRC Press Web site at
http://www.crcpress.com

Contents

Preface

Much fundamental research, technologies, and applications are now moving toward achieving high-precision positioning and higher specifications in production, thus achievable with the requirements of precision motion control up to the order of the submicrometer or nanometer level. This is driven by the emergence of current technologies, such as high-precision manufacturing processes, machines, biotechnology, and nanotechnology. From the early 1980s, semiconductor and biomedical industries demanded high-precision actuators to execute more precise positioning and manufacturing in their processes, and the move toward ever-higher precision has continued to now. Requirements pertaining to the precision of motion vary substantially. As such, high-precision actuators are now in high demand, and are expected to perform various types of movements, from rotation to translation, high torque capability, wide speed range, etc. The application areas of high-precision actuators are diverse in aerospace, microelectronics, biomedical engineering, and nanotechnology.

This book is a result of several years of work in the realization of precise actuators. The primary intent of this book is to report new technologies in the area of precision motion control, which can ultimately be applied in industry. It covers dynamical analysis of precise actuators and strategies of design for various control applications. The book consists of eight chapters treating different topics. The content is suitable for graduate students and engineers in precision engineering.

In what follows, the contents of the book are briefly reviewed.

Chapter 1 introduces the driving forces behind precise actuators, several typical types of the actuators, as well as their applications. Chapter 2 describes nonlinear dynamics of precise actuators and their mathematical forms, including hysteresis, creep, friction, and force ripples.

Chapter 3 presents control strategies for precise actuators based on the Preisach model as well as creep dynamics. The identification algorithm is first proposed to estimate Preisach model parameters, and then the inversion feedforward controller is designed for hysteresis compensation. This strategy is mainly for low frequency. For the case where precise actuators work at high bandwidth relative to the resonant frequencies, another identification and compensation strategy is designed. The proposed methods are illustrated by experimental results.

Chapter 4 develops relay feedback techniques for identifying nonlinearities such as friction and force ripples. By converting the closed-loop system into a multiple-relay feedback system, switching conditions of a stable limit cycle are obtained. Hence, friction and force ripples are identified by numerically solving a set of equations. Simulation and real-time experiments show the practical appeal of the proposed method.

Model predictive control (MPC) has become an attractive feedback strategy, especially for linear systems. MPC solves an online optimization to determine inputs, taking into account the current conditions of the plant, any disturbances affecting operation, and imposed safety and physical constraints. Over the last several decades, MPC technology has reached a mature stage. In Chapter 5, we present an MPC approach based on piecewise affine models that emulate the frictional effects in a precise actuator. Specially, an integral MPC design imposes robustness on a model–plant mismatch near zero speed. Implementation of the real-time control is handled by a gain scheduling table so that the complexity is comparable to the traditional feedforward proportional-integral-derivative (PID).

Chapter 6 presents the concepts of air bearing stages with the corresponding control method. A linear air bearing stage is first considered. Since it is a floating object, eddy current braking is introduced into the system. A nonlinear control with proportional-integral (PI) control is designed for dealing with nonlinear terms. Subsequently, a multi-DOF (degrees of freedom) spherical air bearing stage is presented. An adaptive noise filter and a controller for angular positioning are proposed to achieve high performance. Finally, experimental results are given to show the effectiveness of the proposed control algorithm.

Chapter 7 presents a set of schemes suitable for fault detection and accommodation control of mechanical systems. The basic idea of designing a fault detection scheme is to use the information provided by a model-based nonlinear observer to find failure occurrences. The fault detection decision is carried out by comparing the observer outputs with their signatures. After a fault is detected, the controller is reconfigured by incorporating neural networks that are used to capture the nonlinear characteristics of unknown faults. The designed schemes can achieve the automated fault detection and accommodation control using a dead-zone operator.

Chapter 8 is intended to provide readers with a bridge between the design methods of the previous chapters and their applications. With this purpose in mind, the chapter emphasizes the key issues involved and how to implement the precision motion control tasks in a practical system. These issues are demonstrated by three case studies. The first case study describes a robust adaptive control method for positioning piezoelectric actuators (ultrasonic motor) to achieve highly precise motion. Real-time experimental results are provided to verify the effectiveness of the proposed scheme when applied to high-precision motion trajectory tracking, such as intracytoplasmic sperm injection (ICSI). The second case study is focused on a motion control for a two-dimensional stage that is used to treat a common disease called otitis media with effusion (OME), involving a surgeon inserting a grommet in the eardrum to bypass the Eustachian tube to drain fluid when medication fails. The third application is a vision-based real-time temperature monitoring system, where an object recognition and tracking algorithm will be applied to guide a temperature sensor to monitor the temperature of the working tool while it is carrying out operations.

Tan Kok Kiong
Huang Sunan

MATLAB® is a registered trademark of The MathWorks, Inc. For product information, please contact:

The MathWorks, Inc.
3 Apple Hill Drive
Natick, MA 01760-2098 USA
Tel: 508 647 7000
Fax: 508-647-7001
E-mail: info@mathworks.com
Web: www.mathworks.com <http://www.mathworks.com>

Acknowledgments

The first author is thankful to his PhD students for their research leading to the contents of the book. They include Liu Lei, Chen Silu, Nguyen Hoang Tuan Minh, Liang Wenyu, and Zhang Yi.

The authors thank Dr. Teo Chek Sing and Tan Chee Siong for their assistance in the writing of the book. They also thank the National University of Singapore and Singapore Institute of Manufacturing Technology for their funding support. The editing process would not have been as smooth without the generous assistance of Laurie Schlags, Robin Lloyd-Starkes and Li Ming Leong.

Finally, the authors thank their families for their love and support.

Tan Kok Kiong
Huang Sunan

About the Authors

Tan Kok Kiong earned his B.Eng. in electrical engineering with honors in 1992 and PhD in 1995 from the National University of Singapore.

Dr. Kiong is currently an associate professor with the Department of Electrical and Computer Engineering, National University of Singapore. His current research interests are in the areas of advanced control and autotuning, precision instrumentation and control, and general industrial automation. He has produced more than 160 journal papers to date and has written 5 books, all resulting from research in these areas. He has so far attracted research funding in excess of S$7 million and has won several teaching and research awards.

Huang Sunan earned his PhD degree from Shanghai Jiao Tong University, Shanghai, China, in 1994. Currently, he is a research fellow in the Singapore Institute of Neurotechnology, National University of Singapore. His research interests include error compensation of high-precision machines, adaptive control, neural network control, and rehabilitation robot control.

1

Introduction

In a wide range of industries, many applications require much higher precision over a high bandwidth and speed than traditional actuators can deliver. This increase in demand for higher-precision motion control has led to many new innovations, including high-speed Maglev transportation systems, robotics machines, micromanipulation systems, semiconductors, and the use of piezo-electric materials to create motion. Precise actuators are the motion enablers of the motion control system. They utilize a physical interaction to convert an electrical energy into mechanical motion to achieve high speed and high accuracy resolution. Developments of such precise actuators and their control technologies will have an impact on a wide range of industries, from medical technology to precision tooling machines and 3D printing. The purpose of this chapter is to discuss the drivers of precise actuators, main types of precise actuators, challenges in their control, and precision applications.

1.1 Growing Interest in Precise Actuators

In recent years, electronic control and machine control have become more efficient as new microprocessors, digital signal processors (DSPs), and other electronics chips are providing the control platform with tremendous computing and processing power. Advances in actuators, such as direct drive motors, piezo motors, coil motors, air bearing motors, linear motors, and brushless motors are reducing traditional issues such as backlash, hysteresis, friction, and parasitic system dynamics. The technical field of precision actuators has expanded significantly over the past 30 years to include design method, error compensation, control, actuator, sensor, fault failure, software platform, and design methodology.

Frequently used actuators in the domain of precision and ultraprecision are piezoelectric actuator (PA) and linear motors. For example, PAs have been applied to products such as a piezoelectric buzzer, a printer head, and ultrasonic motors. In precision engineering applications, PAs have been increasingly employed, such as in modern micro- and nanofabrication, dynamic imaging with scanning probe microscopes (SPMs), and advanced spacecrafts with sensitive optical instruments. With the development of ultra-accurate applications, more stringent requirements are presented [1], which lay out the scope of current techniques.

- High bandwidth: In SPMs, the PAs are required to track at very high rates, which may exceed the resonant frequencies of PAs. Currently, most PAs operate at frequencies less than 10% of the resonant frequencies.

- High accuracy: In addition to high-bandwidth requirements, high accuracy is another requirement for PAs. Moreover, both high bandwidth and high accuracy are current requirements in which the precision tracking is required at rates possibly beyond the resonant frequencies.

- Feedforward control: Feedback control is validated at normal working frequency, but it is limited at frequencies higher than the resonant frequencies due to the measurement noise at high frequencies. The model-based inversion feedforward compensation, which relies on the model identification, is a useful technology to increase the tracking rates and enhance the trajectory tracking accuracy at high frequencies, because the feedforward controller is effective for avoiding the measurement noise that is more serious at high frequencies.

The 2009 global market for piezoelectric-operated actuators and motors was estimated to be 6.6 billion, and the market is estimated to reach 12.3 billion by 2014, showing an average annual growth rate of 13.2% per year [2]. Due to the demand from the consumer electronics market beyond computers, hard disk drive (HDD) demand has been experiencing continual growth over the past decade. It was reported in 2007 that 516.2 million hard disk drives were sold [3]. Meanwhile, the linear motor is gaining attention in precision manufacturing. The main driver of linear motor technology is the ever-increasing performance demand in incremental positioning applications [4]. Unlike rotary machines, linear motors require no indirect coupling mechanisms as in gear boxes, chains, and screws coupling. This greatly reduces the effects of contact-type nonlinearities and disturbances, such as backlash and frictional forces. In addition to precise actuator design, the control of precision systems also has a wide range of applications in high-speed and high-accuracy automation. The growth of nanotechnology and nanoscale manufacturing has further raised the positioning requirements for machine and controller design. The performance of such systems depends on the application of advanced feedback control due to the large-scale system of controlled variables and the uncertainties that might affect the system significantly.

1.2 Types of Precise Actuators

In this book, we will focus mainly on two common types of actuation technology used to achieve linear motion: piezoelectric actuator and linear motor.

1.2.1 Piezoelectric Actuator

The piezoelectric actuator (PA) has become an increasingly popular candidate as a precise actuator in industry, due to its ability to achieve high precision and its versatility to be implemented in various applications. More specifically, the PA can provide very precise positioning (of the order of nanometer) and produce forces from low force (a few grams) to high force (up to a few thousand Newtons). This is because PAs have the following appealing properties:

- High bandwidth (at rates of kHz)
- Small displacement (typically several microns)
- High accuracy (typically subnanometer accuracy)
- High force
- Friction-free (flexible joints are commonly used to avoid generating the friction)
- Minimal static energy consumption
- Small size

The increasingly widespread industrial applications of the PA in various optical fiber alignments, mask alignment, and medical micromanipulation surgical robots are self-evident testimonies of the effectiveness of the PA in these application domains. To obtain maximum performance from PAs, various types have been designed.

1.2.1.1 Stack Actuator

The most popular design for PAs is a stack of ceramic layers separated by thin metallic electrodes, called stack actuator (see Figure 1.1). The stack actuator changes its dimension or size when an electric field with a power supply is applied to it. Such change is very small and produces linear motion. This implies that the stack actuator can achieve high-precision displacements (typically 20 nm). The actuator can also produce different forces according to supply voltage. Commercial products for such stack actuators are available from Physik Instrumente, Kinetic Ceramics, and NEC, which can provide precise positioning in microns or at the nanometer level.

1.2.1.2 Piezoelectric Shear Actuator

Piezoelectric shear actuators are also very common, as shown in Figure 1.2. Compared with linear PAs, shear actuators are adapted for small transverse

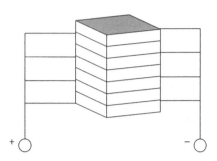

FIGURE 1.1
Piezoelectric stack actuator.

displacements where space is a constraint. They offer very fast response times. The main advantage of the shear actuators is their suitability for a bipolar operational source, whereby the mid-position corresponds to a drive voltage of 0 V. It should be noticed that in a shear actuator, the electric field is applied perpendicular to the polarization direction, which is different from the other types of actuators.

1.2.1.3 Piezoelectric Bending Actuator

In addition to the stack and shear actuators, piezoelectric bending actuators (these actuators are often referred to as benders, piezoelectric cantilevers, or piezoelectric bimorphs) are another important PA. The working principle is that the application of an electric field across the two-layer element produces curvature when one layer expands while the other layer contracts. Typical movement for this kind of actuator is on the micrometer level (from hundreds to thousands of microns), while the bender force generated is small (from tens to hundreds of grams). Figures 1.3 and 1.4 show two common bending configurations that are often used in various applications. The first configuration, as shown in Figure 1.3, is called serial bender and has two piezoelectric layers with two electrodes and an antiparallel polarization connected to each other. The second configuration, as shown in Figure 1.4, is

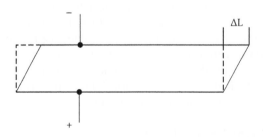

FIGURE 1.2
Piezoelectric shear actuator.

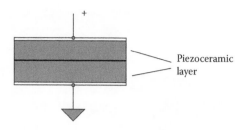

FIGURE 1.3
A series bender.

called parallel bender and has three electrodes, where one of the two surface electrodes is connected to the ground and the other is connected to a voltage, while the middle electrode is connected to another control voltage that varies between zero and a voltage. These three electrodes in the parallel bender can also be connected in such a way that one is attached to each outside electrode and one is attached to the center shim.

1.2.2 Linear Motor

In today's industry, there are a multitude of linear motion control applications. Linear motors are probably the most naturally akin to single-axis or multiple-axis motion applications involving high speed and high accuracy. The main driver of linear motor technology is the ever-increasing performance demand in incremental positioning applications [4]. The main linear motor types available commercially are permanent magnet linear motor, linear piezo motor, and linear air bearing stage.

1.2.2.1 Permanent Magnet Linear Motor

Permanent magnet linear motor (PMLM) is widely used in high-precision motion control. The main driver of requiring PMLM technology is its excellent servo performance. The digital revolution has supported this by allowing great advances in motion controller capabilities and reducing the cost of

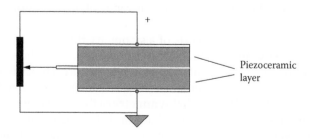

FIGURE 1.4
A parallel bender.

FIGURE 1.5
U-shaped permanent linear motor.

feedback control, hence driving system designers to more often use digital closed-loop control systems. This, in conjunction with advances in magnet materials and power electronics, has caused PMLM technology to emerge as the dominant type of linear motor [4]. The main types of PMLMs are the U-shaped PMLM and tubular PMLM.

The U-shaped PMLM itself is contained within a nonmagnetic steel housing. The device consists of two main elements (see Figure 1.5): a double row of magnets fixed in the steel housing (typically stationary) and a forcer rod containing the motor coils (typically the moving element). As the coil is energized and de-energized, this generates a magnetic field that can drive the forcer rod forward or backward, creating linear motion. In the forcer rod, a circuit board with electronics for the three Hall sensors that determine positioning is hidden, along with the power connection. The U-shaped PMLM offers general cost-effectiveness and a wide range of travel length capabilities for motion control.

The tubular linear motor consists of a stationary thrust rod and a moving thrust block. One example is the LD3810 tubular motor, as shown in Figure 1.6. The thrust rod is a permanent magnet, while the thrust block is an electromagnet winding. This design confers several advantages compared with the other linear motor types by its radial symmetry of the tubular geometry. First, the attractive force between the translator and stator is minimized by such geometry. Second, the linear forces are maximized by the perpendicularity between the circular windings in the thrust block and the magnetic flux pattern. Third, eddy current losses are insignificant due to the slotless design.

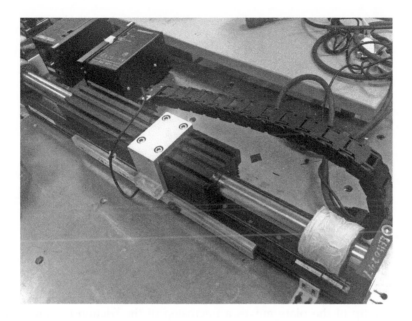

FIGURE 1.6
Tubular permanent linear motor.

Furthermore, the thrust block is designed to serve as a radiator for passive cooling. The installation is simpler by its relatively large allowance of air gap.

1.2.2.2 Linear Piezo Stage

Stack actuators can generate linear movement reaching as stated above. However, the displacement of the piezoelectric stack actuator is still very limited for some applications. One way to solve this problem is to develop mechanical amplifier structures that can increase the displacement of the piezoelectric stack actuator. Such a designed device is called the linear piezo stage. The stage is driven using piezoelectric actuators with two different working principles. The first one is the hybrid transducer actuators or inchworm actuators; it uses separate clamps and drive stages to perform the linear motion [5]. Theoretically speaking, the inchworm piezo stage can produce unlimited linear displacement, but with limited force and response frequency. The main drawback of this type of actuator is that it is difficult to obtain a continuous and smooth motion due to the sequential alternation between the different actuators. The second stage works on the principle of excitation of ultrasonic standing waves within a piezoelectric linear resonator. One example is the M-663 linear piezo stage manufactured by Physik Instrumente, as shown in Figure 1.7. The stage movement depends on the friction generated between the piezo-ceramic plate mounted in the stator and the friction bar attached to the mover. The piezo-ceramic plate is the motor's core piece, which can be excited to produce high-frequency eigenmode oscillations. For each oscillation

FIGURE 1.7
Linear piezo stage.

cycle, the tip of the plate moves a microstep of the friction bar along the guideway. Thus, at the eigenmode frequency, the mover is driven forward or backward through the contact between the tip and the friction part [6]. The minimum incremental displacement of the mover is 0.3 μm. It is measured by a built-in linear encoder with a resolution of 0.1 μm. The maximum push-pull force the M-663 can provide is 2 N, and the velocity can be up to 400 mm/s. Moreover, M-663 is able to self-lock even if it is powered down.

1.2.2.3 Linear Air Bearing Stage

Although both permanent magnetic linear motor and piezo linear stage can provide precision motion control, they still incur friction, which degrades the control precision. Compensation is one way to deal with that, as well as other nonlinearities, such as the ripple forces. Alternatively, air bearing technology can be used when thoroughly smooth motion is needed. It is well known that air bearings have a superior smoothness performance in high-precision motion applications. The linear air bearing stage (see Figure 1.8) is a fully

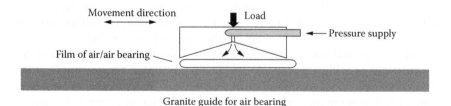

FIGURE 1.8
Working principle of air bearings.

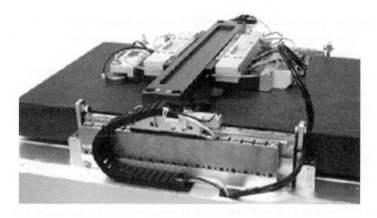

FIGURE 1.9
An air bearing stage developed by Singapore Institute of Manufacturing Technology.

supported air bearing design. Basically, air bearings use pressurized air to provide an air thin film between two objects to support the payloads. The thin-film bearings have no solid-to-solid contact under movement and serve to transfer forces from one to the other. Comparing air bearings with other types of bearings (including rolling bearings and fluid bearings), air bearings have low viscosity and offer lower friction and smaller sensitivity to temperature variation. Meanwhile, they offer a number of advantages to precision positioning systems, such as noncontact, lack of stiction and backlash, and a green operation. Also, noncontact design improves positioning repeatability. Figure 1.9 shows one example of air bearing stage made by Singapore Institute of Manufacturing Technology.

1.3 Applications of Precise Actuators

In precision engineering applications, PAs have been more and more employed, such as modern micro- and nanofabrication, dynamic imaging with scanning tube microscopes (STMs), and advanced spacecrafts with sensitive optical instruments. See the following list for specific applications

Modern micro- and nanofabrication: 3D nanofabrication by femtosecond laser direct writing, nanoscale scratching, submicron lithography, diamond turning machines, etc.

Dynamic imaging with STMs: Three-dimensional nanopatterning and dynamic imaging of molecules, nano-visualization of dynamic biomolecular processes, etc.

Advanced spacecrafts with sensitive optical instruments: Space telescopes, deep-space laser communication, space laser weapons, space interferometers, etc.

Data storage: R/W head testing, spin stands, vibration cancellation, etc.

Life science, medicine, biology: Scanning microscopy, patch clamp, gene manipulation, micromanipulation systems, cell penetration, etc.

Precision metrology: Submicron coordination measuring machines, flatness and roundness measuring systems, image processing, vision and optical inspection, automotive, medical, electronics, optical components, etc.

Precision mechanics and mechanical engineering: Noncircular boring machine, drilling machine, turning machine, wear correction, needle valve actuation, micropumps, knife edge control in extrusion tools, etc.

References

1. S. O. Moheimani. Accurate and fast nanopositioning with piezoelectric tube scanners: Emerging trends and future challenges. *Rev. Sci. Instrum.*, 79, 071101, 2008.
2. iRAP Research Group. Piezoelectric crystals and crystal devices—types, materials, applications, new developments, industry structure and global markets, Electronics Industry Market Research and Knowledge Network. http://www.electronics.ca/publications/products/html.
3. Wolfgang Gruener. Hard drive industry shrugs off economic concerns, posts double digit growth rates. TG Daily. http://www.tgdaily.com/content/view/38533/118/, (accessed January 11, 2009).
4. Michael Backman. Ironcore and ironless linear servo motors—A comparison of linear motor and rotary servo motor systems. Rockwell Automation. http://www.rockwellautomation.com/anorad/products/linearmotors/comparison_wp.html.
5. K. Uchino and J. R. Giniewicz. *Micromechatronics* Series Materials Engineering 22. New York: Marcel Dekker, 2003.
6. Physik Instrumente (PI). M-663 PLiner® linear motor stage. http://www.pi.ws.

2

Nonlinear Dynamics and Modeling

Precise actuators (PAs) have inherent common nonlinear dynamic phenomena associated with mechanical, structural, electrical, and control systems. Studies in these phenomena can increase the understanding of actuator systems and help us improve control quality.

This chapter will describe several typical physical dynamics that appear in precise actuators, such as piezo actuator, shape memory alloy actuator, coil motor, and permanent magnetic motor. There are four common dynamics observed in precise actuators: hysteresis, creep, friction, and ripple force. Their mathematical descriptions will be discussed in this chapter.

2.1 Hysteresis

Hysteresis is a strong nonlinearity that has global memories. It can be observed from many precise actuators, such as piezoelectric actuator, magnetic motor, coil motor, shape memory alloy actuator, etc. Models of hysteresis can be classified into physical models and mathematical models. The physical model is constructed based on physical laws applied to the phenomenon of hysteresis, and thus the model is intuitive to understanding, but it is typically in a complex form that is difficult to identify and use for control purposes. Conversely, the mathematical model is a vehicle to provide an input-output relationship of the actual system, and it is usually more amenable to practical use for identification and control. In the current literature, the ferromagnetic hysteresis model is a physical model, while the Preisach model and the Prandtl–Ishlinskii (P–I) model are two popular mathematical models [1]. The P–I model has been used to compensate the hysteretic nonlinearity of PZT [2]. The Bouc–Wen model has also been applied to describe the nonlinearities of actuators [3], which are akin to the hysteretic characteristics. Compared to the Preisach model, these models are in a highly nonlinear form, possibly including dynamical parameters that are difficult to lay out in a form for identification, and thus impeding their subsequent use for control purposes.

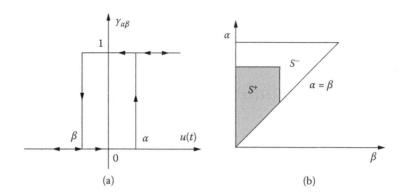

FIGURE 2.1
(a) Preisach relay operator. (b) Preisach triangle.

The hysteresis relay constitutes the basic element of hysteresis in PAs. The outputs of these operators are weighted by the Preisach density function $\mu(\alpha, \beta)$ and then summed continuously over possible values of α and β. The relationship between input voltage and hysteresis output of PAs is represented as [1]

$$f(t) = \iint_{\alpha \geq \beta} \mu(\alpha, \beta)\gamma_{\alpha\beta}[u(t)]d\alpha d\beta \tag{2.1}$$

where $\mu(\alpha, \beta)$ is the density function and $f(t)$ is the hysteresis output. At low frequencies, typically less than 1 Hz, $f(t)$ is approximately equal to the piezo displacement. α and β are the switching threshold values of the hysteresis operator $\gamma_{\alpha\beta}[u(t)]$, as shown in Figure 2.1(a).

In the Preisach model (2.1), the input $u(t)$ is first applied to all the hysteresis operators $\gamma_{\alpha\beta}[u(t)]$. The hysteresis output can thus be considered to be a superposition of a continuous set of two-position relay operators $\gamma_{\alpha\beta}[u(t)]$ over the range of input signal. Let S be a Preisach triangle that is formed by $\alpha \geq \beta$ and the saturation value of input voltages, shown in Figure 2.1(b), and can be divided into S^+ with $\gamma_{\alpha\beta}[u(t)] = \theta_1$ and S^- with $\gamma_{\alpha\beta}[u(t)] = \theta_2$. In PAs with positive input voltage, $\theta_1 = 1$ and $\theta_2 = 0$ can be used.

Three properties of the Preisach model are frequently used, i.e., the rate independence, memory rules, and wiping out properties, which will be employed to construct the identification strategy of the coupling hysteresis and creep.

Rate independence. The rate independence property is represented as [4]

$$\Gamma[u \circ \varphi] = \Gamma[u] \circ \varphi \tag{2.2}$$

where $\circ$ is the composition operator, and φ is the increasing function mapping the considered time onto itself.

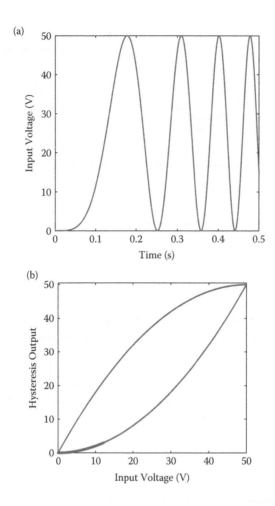

FIGURE 2.2
Illustration of rate dependence of the Preisach model. (a) Input voltage with constant amplitude but variant frequencies. (b) Hysteresis curve of the hysteresis output versus input voltage.

Figure 2.2(a, b) illustrates the rate independence of the Preisach model. The input signal has a constant amplitude but varying frequencies. The resultant hysteresis curve of the hysteresis output versus the input voltage is still invariant.

Remark: The rate independence property indicates that adequate frequencies are not helpful to improve identification of Preisach hysteresis, which is different from the identification of nonhysteretic dynamics where adequate frequencies are necessary to achieve accurate identification.

Memory rules and wiping out property. The details of memory rules of the Preisach model are described in reference [1]. Figure 2.3(a, b) shows the memory rules of the Preisach model. As the input signal $u(t)$ is monotonically

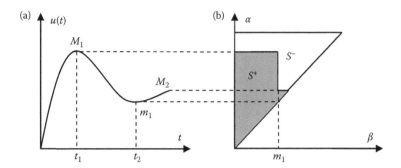

FIGURE 2.3
Illustration of memory rules of the Preisach model. (a) Input signal. (b) Hysteresis representation in the Preisach plane.

increased from zero to the local maximum value M_1, all the hysteresis operators $\gamma_{\alpha\beta}[u(t)]$ with switching values less than M_1 will be activated. Next, the input signal is monotonically decreased from the local maximum value M_1 to the local minimum value m_1, and the $\gamma_{\alpha\beta}[u(t)]$ with switching values larger than m_1 becomes deactivated. Geometrically, this corresponds to a division of the limiting triangle into two regions, i.e., the activated S^+ and the deactivated S^-, as shown in Figure 2.3(b). If $M_2 > M_1$, the extreme M_1 and m_1 will be deleted in the Preisach memory according to the wiping out property. The three properties of Preisach model will be employed to identify the coupling hysteresis and creep dynamics.

2.2 Creep

Another typical phenomenon observed in PAs is the creep, which behaves as a drift. It usually degrades the identification accuracy of the hysteresis model. Figure 2.4 illustrates the observed hysteresis curves in the piezoelectric stage used in the chapter. The hysteresis curves are distorted by the creep. The coupling between the hysteresis and the creep can be observed.

The creep dynamics C can be represented as [5]

$$C = k_c \prod_{i=1}^{m} \frac{z + z_i}{z + p_i} \tag{2.3}$$

where m, z_i, and p_i denote the order, zeros, and poles of the creep model, and k_c is generally coupled with the hysteresis and cannot be separated, except the hysteresis output v can be measured.

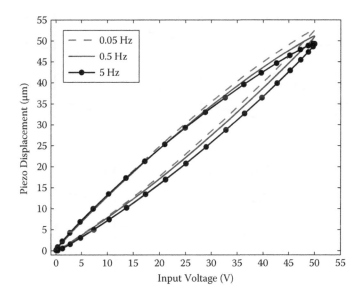

FIGURE 2.4
Hysteresis curves in the experimental piezoelectric stage.

2.3 Friction

The designs and applications of precise motion control systems have also been closely related to investigation of friction between contact surfaces of machine subparts. Friction is the force resisting the relative lateral (tangential) motion of solid service. Humans have tried to make use of friction between two rough stones to start fires since the stone ages. In ancient Egypt, workers learned to put heavy stones on wooden sledges so that easier transportation of these stones was achieved due to much reduced friction during rolling. The usage of friction was even applied to design a complicated loom machine for figure weaving in the silk handicraft industry in the Ming Dynasty of China (1368–1644 A.D.) [6]. The classical understanding of friction was continually investigated in [7,8]; their findings are summarized in the following laws:

- The force of friction is directly proportional to the applied load.
- The force of friction is independent of the apparent area of contact.
- Kinetic friction is independent of sliding velocity.

The laws above describe the behaviors of so-called Coulomb friction. The relay-type Coulomb friction model in Figure 2.5(a) is a simple but efficient one to describe friction behavior when the motion is nonstuck and in medium speed. In order to describe friction behavior under different moving

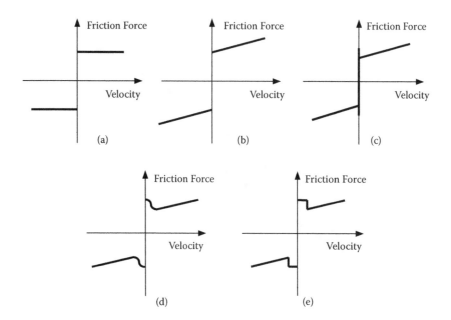

FIGURE 2.5
Various friction models. (a) Coulomb. (b) Coulomb + viscous. (c) Static + Coulomb + viscous.
(d) Negative viscous + Coulomb + viscous: form A. (e) Negative viscous + Coulomb + viscous:
form B.

conditions, various other friction models are developed. For the motion with
high speed, the viscous friction needs to be considered; thus the Coulomb
+ viscous friction model is set up, as shown in Figure 2.5(b). For motion involving stiction behavior, the ideal of static friction is introduced, forming the
most commonly used friction model in engineering: the stiction + Coulomb +
viscous friction model [9], as shown in Figure 2.5(c). For example, this friction
model is used for setting up a physical model of a valve positioning system,
where strong skip-slip behavior is observed.

The increasing demands on the precision engineering boost the modeling
of friction in a more accurate way. As the machine accelerates from zero velocity, the friction will first drop from maximum static friction to Coulomb
friction, and then increase due to the viscosity, forming the negative viscous +
Coulomb + viscous friction model [10], as shown in Figure 2.5(d). This model
may be approximated as a four-parameter segmental hard-nonlinearity-type
model [11], as shown in Figure 2.5(e), which is easier for further analysis by
decomposition. The above models approximate the friction force as a function of steady-state velocity. The mathematical expressions of Figure 2.5 are
discussed below.

The Coulomb friction model is given by

$$F = F_c sgn(\dot{x}) \tag{2.4}$$

where F_c is the Coulomb friction, sgn is the sign function, and $\dot{x}$ is the motion velocity. For the viscous friction, it is normally described as

$$F = F_v \dot{x} \tag{2.5}$$

where F_v is the viscous friction. The mathematical expression of Figure 2.5(b) is given by combining the viscous friction together with Coulomb friction. For the expression of Figure 2.5(c), it is necessary to model the station, which has to be modeled by using an external force F_e, that is,

$$F = \begin{cases} F_e & \text{if } \dot{x} = 0 \text{ and } |F_e| < F_s \\ F_s sgn(F_e) & \text{if } \dot{x} = 0 \text{ and } |F_e| \geq F_s \end{cases} \tag{2.6}$$

where F_s is the breakaway force. Thus, the expression of Figure 2.5(c) is given by combining the stiction together with the viscous friction and Coulomb friction. The expression of Figure 2.5 is given by introducing the Stribeck friction

$$F = F_c sgn(\dot{x}) + (F_s - F_c)e^{-|\dot{x}/\dot{x}_s|^\delta} + F_v \dot{x} \tag{2.7}$$

where $\dot{x}_s$ is the Stribeck velocity that defines the region in which such an effect is present, and $\delta > 0$. Figure 2.5(d) is an approximation of Figure 2.5(c) such that it can be described by a relay-type method. Chapter 4 will discuss this approximation in detail.

2.4 Force Ripples

Permanent magnet linear motors (PMLMs) are now widely used in the precision manufacturing industries since, among the electric motor drives, they are probably the most suitable choice for applications involving high-speed, high-precision motion control. PMLM is designed by incorporating rare permanent magnets so that it is able to develop much higher flux without significant heating. However, permanent magnets also introduce some extra nonlinear dynamics. The major nonlinear phenomena faced by a PMLM are the force ripple and friction. Force ripples are strong, position-dependent forces arising from the magnetic structure of a PMLM. The two primary components of the force ripple are the cogging (or detent) force and the reluctance force. The cogging force arises as a result of the mutual attraction between the magnets and iron cores of the translator [12]. Notice that this force exists even in the absence of any winding current, and it exhibits a periodic relationship with respect to the position of the translator relative to the magnets. The reluctance force is due to the variation of self-inductance of the windings with respect to the relative position between the translator and the magnets. Thus, it also has a periodic relationship with the translator–magnet position.

Force Ripple Under Constant Velocity Mode

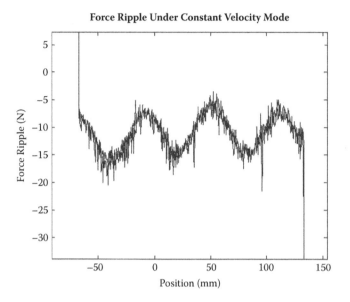

FIGURE 2.6
Open-loop response of a linear permanent magnet motor.

Force ripple is highly undesirable in motion control, since it will create bumps along the direction of motion. Let us see this phenomenon by an experiment. Figure 2.6 shows a real-time step response of a linear permanent magnet motor under constant velocity mode. The bumps are observed clearly in this figure. Additionally, frictional force arises from the contact between the translator and the track [13]. The limits cycle oscillation induced by friction causes small tracking errors in steady states, and it also limits the achievable closed-loop bandwidth [14]. Through alternate mechanical and material design, force ripples and friction may be kept to tolerable levels, but these approaches can be expensive and compromise on other specifications. An alternate approach is to suppress these nonlinear effects through the control system. This requires us to build a model of the force ripple. From Figure 2.6, it is observed that the force ripple is related to the position, and it can be represented by a sinusoidal-type signal. Thus, a first-order model of the force ripple is given by a periodic signal:

$$F = A sin(\omega x + \phi) \tag{2.8}$$

where A is the amplitude of the force ripple, ω is the periodic frequency, x is the position of the motion, and ϕ is the initial phase.

References

1. I. Mayergoyz. *Mathematical models of hysteresis*. Berlin: Springer-Verlag. 2003.
2. K. Kuhnen. Modeling, identification and compensation of complex hysteretic nonlinearities: A modified Prandtl-Ishlinkskii approach. *European Journal of Control*, 9(4), 407–418, 2003.
3. J. C. Lin and S. R. Yang. Precise positioning of piezo-actuated stages using hysteresis observer based control. *Mechatronics*, 16(7), 417–426, 2006.
4. A. Visintin. *Differential models of hysteresis*. New York: Springer-Verlag, 1996.
5. L. E. Malvern. *Introduction to the mechanics of a continuous medium*. Englewood Cliffs, NJ: Prentice-Hall, 1969.
6. Y. Song. The exploitation of the work of nature. University Park: Pennsylvania State University Press, 1966. First print in 1637 in Chinese, English translation by E. Z. Sun and S. C. Sun.
7. L. Da Vinci, J. P. Richter, and R. C. Bell. *The notebooks of Leonardo da Vinci*. Dover, NY: Dover Publications, 1970.
8. G. Amontons. On the resistance originating in machines. In *Proceedings of the French Royal Academy of Sciences*, 1699, pp. 206–222.
9. A. J. Morin. New friction experiments carried out at Metz in 1831–1833. In *Proceedings of the French Royal Academy of Sciences*, 1833, vol. 4, pp. 1–128.
10. L. C. Bo and D. Pavelescu. The friction-speed relation and its influence on the critical velocity of the stick-slip motion. *Wear*, 82(3), 277–289, 1982.
11. M. S. Kim and S. C. Chung. Friction identification of ball-screw driven servomechanisms through limit cycle analysis. *Mechatronics*, 16, 131–140, 2006.
12. I. Boldea and S. A. Nasar. *Linear motion electromagnetic devices*. chap. 2–4. NY: Taylor & Francis, 2001.
13. B. Armstrong-Helouvry, P. Dupont, and C. C. deWit. A survey of models, analysis tools and compensation methods for control of machines with friction. *Automatica*, 30(7), 1083–1138, 1994.
14. H. Olsson and K. J. Astrom. Friction generated limit cycles. *IEEE Transactions on Control Systems Technology*, 9(4), 629–631, 2001.

3

Identification and Compensation of Preisach Hysteresis in Piezoelectric Actuators

Hysteresis contributes to the main uncertainty, which affects the control performance, among the nonlinearities present in piezoelectric actuation systems. In the open loop, the maximum error from hysteresis is 10–15% of the total displacement of piezoelectric actuators (PAs) [1]. This error may not be tolerable for precision applications. The modeling and identification of the hysteresis nonlinearity in PAs can enhance the control performance and the identification accuracy of nonhysteretic dynamics at higher frequencies. Currently, to avoid the hysteresis effect, only 5% of the travel range of PAs is used to identify the transfer functions [2]. Moreover, an input voltage slower than 1 Hz is typically not used for identification due to the hysteresis non-linearity. Thus, the identification issue of hysteresis will be discussed first in this chapter.

PAs are typically quasi-static at low frequencies [3], and can be represented by a rate-independent hysteresis model. Generally, hysteresis models can be classified into mathematical models and physical models [4–6]. The mathematical model is a vehicle to provide an input-output relationship of the actual system, and it is usually more amenable to practical use for identification and control. Conversely, the physical model is constructed based on physical laws applied to the phenomenon of hysteresis, and thus the model is intuitive to understanding, but it is typically in a complex form that is difficult to be identified and not often used for control purposes [7]. In the current literature, examples of mathematical models are the Preisach model and the Prandtl–Ishlinskii (P–I) model [8,9]. The P–I model is a special case of the Preisach model [10].

The Preisach model is popular and effective to describe the quasi-static hysteresis of PAs, but the accurate identification of the Preisach model is still not solved well due to the large number of split lattices and the corresponding density values. Preisach hysteresis is a static nonlinearity and has global memories. All the extreme values of input history can affect current and future outputs. The classical Preisach model satisfies the wiping out and the congruency properties [6]. Other mathematical models, such as the Bouc–Wen model [11,12], have also been applied to describe hysteresis behaviors of PAs.

Compared to the Preisach model, these models are in a highly nonlinear form, possibly including dynamical parameters that are difficult to be laid out in a form for parameter identification and model-based inversion compensation. Modeling of the Preisach hysteresis entails essentially the identification of Preisach density functions. Hu and Song identified these functions by differentiating the measurements [13,14], possibly causing the identified functions to be sensitive to measurement noise. Tan and Iyer developed recursive schemes for parameter identification and designed a closest-match algorithm for the compensation of the Preisach hysteresis [15,16]. Henze provided approaches for the identification of the Preisach function based on different distribution characteristics [17], but these approaches rely on assumptions of the form of the density functions and are typically not amenable to be used for the purpose of producing a model for hysteresis compensation. As evident in the published literature, approximate Preisach density functions can be identified more efficiently through a discretized Preisach plane by transforming the double integral of density functions to a numerical summation, thus reducing the effort to an identification of a set of finite parameters in the linear regression form. Furthermore, no restrictive assumption on the density functions is necessary. However, to achieve an accurate and smooth approximation of these functions from a discretized plane, a large number of the split lattices will be needed, and equivalently, a large number of model parameters is to be determined from the data. This leads to a requirement to collect a large amount of sufficiently exciting data to satisfy a persistent excitation (PE) condition for parameter estimation [16,18]. Under practical conditions, this issue is equivalent to solving an ill-conditioned inverse problem [19].

Various hysteresis compensation approaches have also been investigated. A high-gain feedback controller in [20] is presented to suppress hysteresis, but the achieved bandwidth with adequate tracking accuracy is significantly decreased. The compensation of hysteresis through model-based inversion is adopted in some cases [21]. For example, an adaptive controller with a parameterized inverse hysteresis is designed in [22], but the model is too simple to describe global memories of PAs; an adaptive technique without requiring an inversion of the hysteresis is proposed in [23], but the density function is still assumed in the controller.

This chapter presents a Preisach-based inversion feedforward controller without the measurement of hysteresis output in real time. The least-squares estimation algorithm and the revision of identification are first proposed, both based on singular value decomposition (SVD). The identified model is used to design a feedback-feedforward controller for hysteresis compensation and motion tracking. Experimental studies on a piezoelectric stage are provided to demonstrate the performance of the identification and compensation strategy. However, these results are obtained at low frequencies. The second part of this chapter presents an identification technique and a model-based composite controller to deal with the case where PA works at high frequencies.

3.1 SVD-Based Identification and Compensation of Preisach Hysteresis

3.1.1 Parameter Identification of Preisach Hysteresis

Identification of the Preisach model essentially entails the identification of the density functions. In the continuous form, the parameter $\mu(\alpha, \beta)$ is continuous over the limiting region, and it is thus difficult to identify the continuous Preisach functions. In this section, the limiting region will be considered to comprise discrete lattices. Each lattice cell may have a weight assigned to it that is the discrete equivalent of a specific lattice density. The hysteresis output can be computed by transforming the double integral to a numerical summation as shown in Equation (3.1), which is also linear in parameter and suitable for estimation and control.

$$f(k) = \sum_{i=1}^{L} \sum_{j=1}^{i} \mu_{ij} \gamma_{ij}[u(k)]s_{ij} \tag{3.1}$$

where $f(k)$ and $u(k)$ are the hysteresis output and input voltage at time instant k, respectively, and s_{ij} denotes the area of lattice (i, j). The Preisach plane is discretized into $L \times L$ lattices. Thus effectively with the symmetry, there are $L(L+1)/2$ lattices in the Preisach triangle to be identified.

Let $v_{ij} = \mu_{ij}s_{ij}$. Moreover, s_{ij} is known. The $f(k)$ at time k is rewritten as

$$f(k) = A_k X \tag{3.2}$$

where $A_k = [\gamma_{11}(k) \quad \gamma_{21}(k) \quad \gamma_{22}(k) \quad \gamma_{31}(k) \cdots \gamma_{LL}(k)]$, $X^T = \begin{bmatrix} v_{11} & v_{21} & v_{22} & v_{31} \cdots v_{LL} \end{bmatrix}$, X is to be estimated.

3.1.1.1 Least-Squares Estimation by SVD

The dimension of X is large and the PE condition may not be satisfied; thus least-squares estimation using SVD is employed in this section. Over a time range of $t_1 < t < t_N$, the data samples are collected at time instances $t_1, \cdots, t_i, \cdots, t_N$. With N samples, Equation (3.36) can be produced and posed in the following matrix form:

$$AX = Y \tag{3.3}$$

where $A = [A_1^T \ A_2^T \cdots A_N^T]^T$, $Y = [f(1) \ f(2) \cdots f(N)]^T$, X and Y belong to norm linear spaces, and A is a matrix mapping X to Y. If $M = A^T A$ is nonsingular, the least-squares estimation of μ is unique and given by [18]

$$\hat{X} = M^{-1}A^T Y \tag{3.4}$$

where $\hat{X}$ is the estimation of X.

If $A^T A$ is singular, there will be infinite solutions. To identify the Preisach parameters, detailed discretization is needed. If the discretization level L is 60, the sampling time is 120 s, and the sampling interval is 1 ms, the dimension of matrix A is 120,000 × 1830, and it is difficult to compute $A^T A$. Thus, iterations are employed to compute M and $A^T Y$.

First, to allow the least-squares solution, Equation (3.5) is formed.

$$\left(A_1^T A_1 + A_2^T A_2 + \cdots + A_N^T A_N\right) X = A_1^T f(1) + A_2^T f(2) + \cdots + A_N^T f(N) \quad (3.5)$$

The iterations to compute M and $A^T Y$ are shown as follows:

$$\begin{cases} M^{(k)} = M^{(k-1)} + A_k^T A_k \\ (A^T Y)^{(k)} = (A^T Y)^{(k-1)} + A_k^T f(k) \end{cases} \quad (3.6)$$

where $k = 1, 2, \cdots, N$. $M^{(k)}$ is the matrix M at time k, $M^{(0)} = 0$, and $Y_A^{(0)} = 0$. If $A^T A$ is singular or significantly ill-conditioned, the PE condition is not satisfied, which is likely to occur in this application of Preisach identification. In this case, the SVD approach is used to obtain the pseudoinverse. The SVD of $A^T A$ is given by

$$M = U \Sigma V \quad (3.7)$$

where

$$U = [u_1 \quad u_2 \cdots u_n]$$
$$V = [v_1 \quad v_2 \cdots v_n]$$
$$\Sigma = \text{diag}([\sigma_1, \sigma_2, \cdots, \sigma_n])$$

Singular values $\sigma_1 \geq \sigma_2 \geq \cdots \geq \sigma_n$, U, and V are unitary matrices.

Assume the rank of matrix M is k, then $\sigma_{k+1}, \sigma_{k+2}, \cdots, \sigma_n = 0$. If $\sigma_j/\sigma_1 \ll 1$, $j = r+1, r+2, \cdots, k$, M is ill-conditioned and can be decomposed as follows:

$$M = \sum_1^r \sigma_i u_i v_i^T + \sum_{r+1}^k \sigma_i u_i v_i^T \quad (3.8)$$

Small singular values can be truncated to yield improved least-squares estimation in an ill-conditioned situation. The approximation of the pseudoinverse of A can be given by

$$A^+ \approx V_r \Sigma_r^{-1} U_r^T A^T \quad (3.9)$$

where A^+ denotes the pseudoinverse of A and

$$\Sigma_r^{-1} = \text{diag}\,[1/\sigma_1 \ 1/\sigma_2 \cdots 1/\sigma_r]$$

The pseudoinverse of A is reformulated as

$$A^+ \approx \sum_1^r \frac{1}{\sigma_i} v_i u_i^T A^T \tag{3.10}$$

Finally, the estimation of X in least-squares sense is given by

$$\hat{X} = A^+ Y \tag{3.11}$$

where $\hat{X}$ is the estimation of X.

3.1.1.2 Identification Revision Using SVD Updating

In this part, the Preisach identification in (3.11) is revised using SVD updating, since Preisach estimation by SVD in least-squares sense is time-consuming. The initial identified result is used and the density values are revised according to new data. Bunch and Nielsen provided some methods of SVD revision and updating [24]. Brand applied a rank 1 modification to movie recommender systems [25]. In this section, the identification revision is based on the initial values A_0 and Y_0 in the previous section. For A_0 and Y_0, the following equation exists:

$$A_0 X = Y_0 \tag{3.12}$$

When a new vector a_1 and scalar output y_1 are added to the matrix, the following equation is achieved:

$$\begin{bmatrix} A_0 \\ a_1 \end{bmatrix}^T \begin{bmatrix} A_0 \\ a_1 \end{bmatrix} X = \begin{bmatrix} A_0 \\ a_1 \end{bmatrix}^T \begin{bmatrix} Y_0 \\ y_1 \end{bmatrix} \tag{3.13}$$

where Y_0 is a vector and y_1 is a scalar.

Let M_0 and M_1 denote $A_0^T A_0$ and $A_0^T A_0 + a_1^T 1_1$, respectively. Using the analysis in Section 3.1.1.1, the initial estimation is written as $M_0 = U_r \Sigma_r V_r^T + \varepsilon_r$, ε_r is the residual error. Then, M_1 can be written as

$$M_1 = U_r \Sigma_r V_r^T + a_1^T a_1 + \varepsilon_r \tag{3.14}$$

Neglecting ε_r, M_1 can be reformulated as

$$M_1 \approx \begin{bmatrix} U_r & a_1 \end{bmatrix} \begin{bmatrix} \Sigma_r & 0 \\ 0 & I \end{bmatrix} \begin{bmatrix} V_r & a_1 \end{bmatrix}^T \tag{3.15}$$

The components of a_1 orthogonal to the space spanned by U_r and V_r are given by

$$\begin{cases} p = a_1 - U_r U_r^T a_1 \\ q = a_1 - V_r V_r^T a_1 \end{cases} \tag{3.16}$$

Let $R_U = \|p\|$ and $R_V = \|q\|$, then the unit vector u_p and v_q can be written as

$$\begin{cases} u_p = p/R_U \\ v_q = q/R_V \end{cases} \tag{3.17}$$

According to the method provided in [25], $[U_r \quad a_1]$ and $[V_r \quad a_1]$ can be written as

$$[U_r \quad a_1] = [U_r \quad u_p] \begin{bmatrix} I & U_r^T a_1 \\ 0 & R_U \end{bmatrix} \tag{3.18}$$

$$[V_r \quad a_1] = [V_r \quad v_q] \begin{bmatrix} I & V_r^T a_1 \\ 0 & R_V \end{bmatrix} \tag{3.19}$$

Then M_1 can be written as

$$M_1 = [U_r \quad u_p] K [V_r \quad v_q]^T \tag{3.20}$$

where

$$K = \begin{bmatrix} I & U_r^T a_1 \\ 0 & R_U \end{bmatrix} \begin{bmatrix} \Sigma_r & 0 \\ 0 & I \end{bmatrix} \begin{bmatrix} I & V_r^T a_1 \\ 0 & R_V \end{bmatrix}^T$$

The SVD of M_1 is transformed to the SVD of K, which is $(r+1) \times (r+1)$, as is shown in Equation (3.21). The SVD of the small matrix K can save computing time, compared to the SVD of the large matrix M_1.

$$K = U_K \Sigma_K V_K^T \tag{3.21}$$

where $U_K^T U_K = I$ and $V_K^T V_K = I$.

Then, the approximation SVD of M_1 can be achieved as

$$M_1 = ([U_r \quad u_p] U_K) \Sigma_K ([V_r \quad v_q] V_K)^T \tag{3.22}$$

Let U_1 be the $1 : r$ columns of $[U_r \quad u_p] U_K$, V_1 be the $1 : r$ columns of $[V_r \quad v_q] V_K$, and $\Sigma_1 = \Sigma_K(1 : r, 1 : r)$. The SVD updating with the fixed rank r is obtained as

$$M_1 = U_1 \Sigma_1 V_1^T \tag{3.23}$$

Finally, the estimation revision of X is given by

$$\hat{X} = V_1 \Sigma_1^{-1} U_1^T A^T Y \tag{3.24}$$

where $\hat{X}$ denotes the estimation of X.

3.1.1.3 Simulation Study of Proposed Identification Approach

This section presents the simulation study of the proposed hysteresis identification and its revision using SVD. Additionally, a measure of the estimation error of the Preisach density is defined as

$$
\begin{cases}
\|\mu - \hat{\mu}\| = \sqrt{\sum_{j=1}^{L} \sum_{i=1}^{j} (\mu_{ij} - \hat{\mu}_{ij})^2} \\
\|\mu\| = \sqrt{\sum_{j=1}^{L} \sum_{i=1}^{j} \mu_{ij}^2}
\end{cases}
\tag{3.25}
$$

where $\hat{\mu}$ is the estimation of μ and can be directly computed using $\hat{X}$ and s_{ij}.

For comparison, the parameter vector X is also estimated using the projection algorithm in [18].

$$
\hat{X}_k = \hat{X}_{k-1} - \gamma_L \frac{(\hat{f}(k) - f(k-1)) A_{k-1}}{\theta + A_{k-1}^T A_{k-1}}
\tag{3.26}
$$

where $0 < \gamma_L < 1$, $\theta > 0$, and $\hat{X}_k$ is the estimation of X at the time instant k.

Let $\alpha_{\max} = 10$, $\alpha_{\min} = 0$, $\beta_{\max} = 10$, $\beta_{\min} = 0$, $\mu(\alpha, \beta) = 4$, and $L = 50$. A sinusoidal input signal $10\sin(5t)$ is used. Moreover, the white noise with root mean square (RMS) of 0.01 μm is added to the measurement. The sampling interval is 0.001 s and the number of sampling points is 4000. Figure 3.1(a) shows the estimated error using the projection algorithm. The estimated error $\|\mu - \hat{\mu}\|$ is 128.3, and the relative error $\|\mu - \hat{\mu}\|/\|\mu\|$ is 91.6%. The parameter identification error of the projection algorithm is significant. Figure 3.1(b) shows the estimated error using the SVD-based identification.

The dimension of A is 4000×1275, and we can get the determinant $\det(M) = 0$. The rank of M is 99. Then, r is set to 64 with $\sigma_{r+1}/\sigma_1 = 0.001$; the estimation error $\|\mu - \hat{\mu}\|$ and relative error $\|\mu - \hat{\mu}\|/\|\mu\|$ of the SVD-based identification are reduced to 6.67 and 4.7%, respectively. The difference between the real and estimated $\mu(\alpha, \beta)$ is small.

Finally, with SVD updating using 50 new points, estimation error $\|\mu - \hat{\mu}\|$ is reduced from 6.67 to 5.9, as shown in Figure 3.2. It can be seen that the SVD updating improves the estimation performance. The SVD-based approach gives better identification of Preisach hysteresis. The relative strengths of the identification approach when applied for this purpose will be investigated and highlighted through experimental studies.

3.1.2 Compensation Strategy of Preisach Hysteresis

3.1.2.1 Preisach-Based Inversion Compensation

This section presents a Preisach-based inversion to compensate the hysteresis in PAs. The Preisach-based inversion is rate independent because of the rate independence of the Preisach model [10], which simplifies the hysteresis compensation. The model-based inversion can be computed offline based on the identified Preisach model and the reference trajectories.

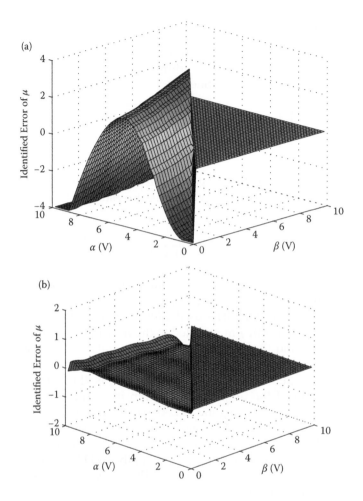

FIGURE 3.1
Estimated error of density function. (a) Projection algorithm. (b) SVD-based least squares.

First, with the models obtained via the possible approaches presented in Section 3.1.1, a hysteresis compensator based on these models can be designed. Figure 3.3 shows the flowchart of a Preisach-based inversion compensator. At time instant k, define the reference displacement as $x_r(k)$, the estimated hysteresis output as $\hat{f}(k)$, and the control action as $u_{ff}(k)$. The feedforward compensator works to obtain $u_{ff}(k+1)$ based on the identified Preisach hysteresis $\hat{H}$. Figure 3.4 shows the Preisach-based inversion feedforward controller. $x_r, d,$ and u_{ff} denote the reference trajectory, the output disturbance, and the feedforward control signal, respectively. From the reference trajectory and its memory curve, the feedforward compensator will compute the feedforward control signal based on the identified density function, and

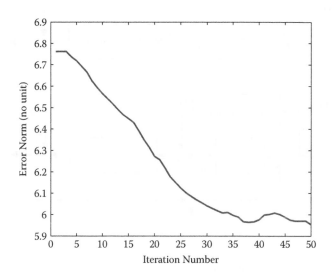

FIGURE 3.2
Converge error of SVD updating.

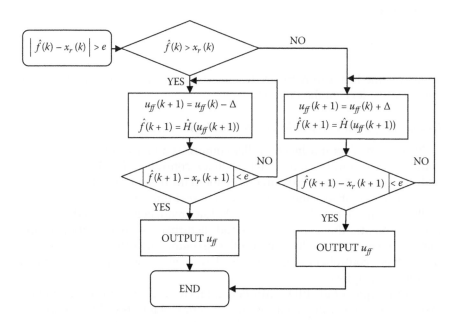

FIGURE 3.3
Flowchart of Preisach-based inversion compensator.

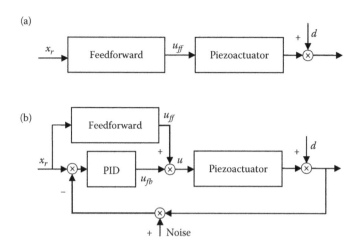

FIGURE 3.4
(a) Feedforward control. (b) Composite control.

enforce the output of the PA to track the desired trajectory, thereby compensating the effects of the hysteresis.

In the flowchart, e is the tracking error bound, and Δ is the iteration step and given as

$$\Delta = \lambda \frac{u_{\max}}{L} \tag{3.27}$$

where λ is the factor to regulate the step, which is less than 1, and L is the discretization level of the Preisach plane.

3.1.2.2 *Proposed Composite Control Strategy*

In this section, a composite controller comprises a Preisach-based inversion feedforward controller and a PID feedback controller. The Preisach-based feedforward can be used to compensate the static hysteresis in PAs, but commonly there are offsets and disturbances in real-time control of PAs; thus, a feedback controller is also necessary. At low frequencies, the feedback controller eliminates the residual errors and disturbances. As the reference frequency increases, the feedback controller also suppresses the dynamic effects. The feedback controller design is not the key issue in this chapter. Thus, a PID controller is employed.

The techniques of the PID controller and its tuning are mature. Theoretical analysis, such as stability analysis, and experimental tests of PID controllers have been applied in piezo systems [14,26], but the tracking performance is still limited due to the hysteresis effect. Based on accurate identification of hysteresis, this chapter uses Preisach-based inversion feedforward to enhance the tracking performance. Figure 3.4(b) shows the composite control of the

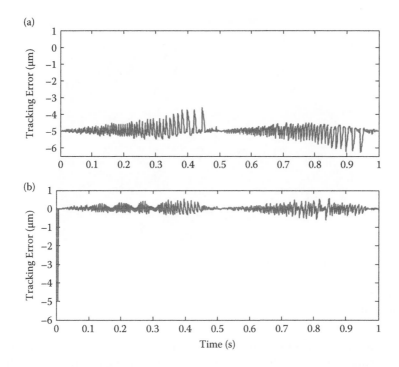

FIGURE 3.5
Tracking errors. (a) Preisach-based inversion feedforward control. (b) Composite control.

PAs. u_{fb} denotes the feedback control signal. The measurement noise is also considered.

3.1.2.3 Simulation Study of Proposed Compensation Strategy

The proposed compensation strategy is simulated in this section. The Preisach hysteresis and the estimated results in Section 3.1.1.3 are used. Moreover, the output disturbance of 5 μm is also added to the system. The reference signal is given as $20(1 - \cos 2\pi t)$. The performance of Preisach-based inversion feedforward is shown in Figure 3.5(a). The RMS tracking error is 5.02 μm. However, the output disturbance is not suppressed by the feedforward controller. Conversely, the RMS tracking error is reduced to 0.22 μm with the proposed composite controller, while the proportional and integral gains of the PID controller are set to 0.001 and 3, respectively. The constant disturbance is also suppressed, as shown in Figure 3.5(b).

3.1.3 Experimental Studies

This section presents the experimental studies of the proposed identification and compensation approaches. First, to validate the SVD-based least-squares

Piezoelectric Stage

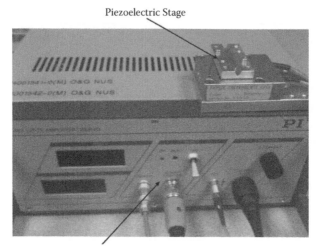

Sensor Conditioning and Amplifier

FIGURE 3.6
Experimental setup.

estimation of Preisach hysteresis, the Preisach density function is identified at low frequencies where the piezo displacement can be regarded as the hysteresis output. The proposed hysteresis compensation strategy is also verified. This section proposes the experimental study of the Preisach hysteresis identification. It also provides the Preisach-based inversion feedforward and the composite control at low frequencies. To extend the application of the identified Preisach hysteresis, this section presents the proposed composite control at high frequencies where the hysteresis output is not measurable due to the dynamics effect. The Preisach-based inversion feedforward is computed according to the reference trajectory and identified Preisach model.

3.1.3.1 Experimental Setup

The experimental setup consists of a piezoelectric stage, a linear voltage amplifier, a linear variable differential transformer (LVDT) displacement sensor, and a dSPACE 1104 board. Figure 3.6 illustrates the piezoelectric stage, amplifier, and sensor conditioning. The travel of the stage is 100 μm. The amplifier has an output voltage range of $[-20, 120]$ V. The noise in the measurement signal is white noise with the root mean square (RMS) value of 0.009 μm.

3.1.3.2 Hysteresis Identification at Low Frequencies

The hysteresis identification of the piezoelectric stage is finished at frequencies lower than 1 Hz. The Preisach model is still used to represent the quasi-static hysteresis. At low frequencies, the hysteresis of a PA is quasi-static [3], since the vibration and electric dynamics approach a DC gain [27]. Therefore, the low-frequency piezo displacement without drift can be regarded as the hysteresis output [28]. To avoid the high-frequency dynamics, the Preisach hysteresis of the piezoelectric stage is identified using smooth input voltage at low frequencies such that the quasi-static assumption holds. The input voltage range is set to [0, 60] V, i.e., the parameters $\alpha_{\min} = 0$, $\beta_{\max} = 60$, $\beta_{\min} = 0$, and $\beta_{\max} = 60$. The input signal is constructed as follows:

$$u(t) = \left[1 + \frac{t - 120 \lfloor t/120 \rfloor}{2} \right] \frac{1 - \cos \pi t}{2} \tag{3.28}$$

where $\lfloor \cdot \rfloor$ is the *floor* function, which rounds elements to their negative integers.

Moreover, the sampling interval is set to 1 ms and 120,000 points are sampled. Using the iterative method in Equation (3.6), the matrix $M = A^T A$ and $A^T Y$ are achieved. Figure 3.7(a) shows the singular values of matrix M. The maximum singular value is 2.541×10^7, and the nonzero minimum singular value is 0.734. The condition number of M is 3.462×10^7. Thus, the matrix M is significantly ill-conditioned. The singular values that are less than 2.5 are truncated. Figure 3.7(b) shows the corresponding identification result of the Preisach density function. The identified density values are positive and their inversion is easy to achieve.

Two methods are used to test the soundness of the identified hysteresis. First, this section presents the comparison of hysteresis curves according to the measured and simulated data. Then, the Preisach-based inversion feedforward in Section 3.1.2.1 is used to test the identification soundness, since the model-based inversion is sensitive to modeling error. Figure 3.8(a) shows the measured and simulated hysteresis curves at 1 Hz. The two curves are close to each other. Figure 3.8(b) shows the curve of the actual piezo displacement versus the desired displacement at 1 Hz. The hysteresis curve is significantly reduced. Thus, the hysteresis identification is satisfactory.

3.1.3.3 *Performance of Proposed Composite Controller at Low Frequencies*

In this section, the proposed composite control strategy is validated at low frequencies where the piezo output can be regarded as the hysteresis output. The inverse Preisach feedforward is given in Figure 3.3. Moreover, the parameters are $\lambda = 0.1$, $e = 0.2$, $L = 60$, and $u_{\max} = 60$ V. Equation (3.29) shows the harmonic reference trajectory x_{r1} with a large amplitude.

$$x_{r1} = 20(1 + \sin 2\pi t) \tag{3.29}$$

Additionally, the proportional–integral–derivative (PID) feedback controller is designed using Ziegler–Nichols rules. A relay is employed to obtain the ultimate gain and ultimate period. Applying the tuning method, the PID

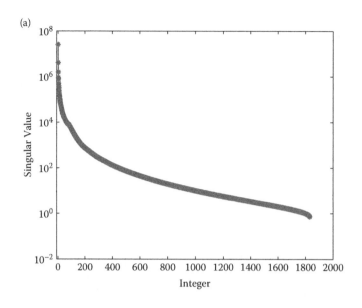

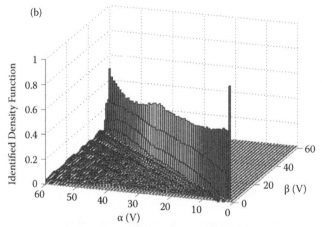

FIGURE 3.7
(a) Singular values. (b) Identified density function.

controller K_{fb} is given by

$$K_{fb} = 0.6K_p \left(1 + \frac{2}{T_i s} + \frac{T_i}{8} \frac{s}{\epsilon s + 1} \right) \tag{3.30}$$

where the ultimate gain K_p is 2.62, the ultimate period T_i is 0.002 s, ϵ is a small positive value to reduce the bandwidth of the derivative, and ϵ is set to 0.005.

To test whether the PID controller in Equation (3.30) achieves its limit, K_p is increased. When the PID gain K_p is increased by 8%, significant chattering

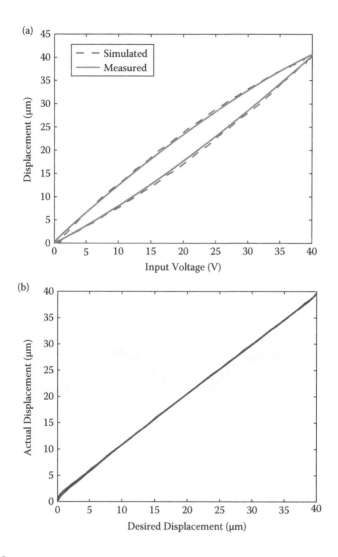

FIGURE 3.8

Validation of hysteresis identification. (a) Comparison of the measured and simulated hysteresis curves at 1 Hz. (b) Actual displacement versus desired displacement at 1 Hz.

arises and the tracking performance worsens. It indicates that the PID controller achieves its limit.

To measure the tracking performance, the percent root mean square (RMS) error e_{rms} as a percentage of the output range is defined as [2]

$$e_{rms}(\%) = \left(\frac{\sqrt{\frac{1}{n} \sum_{i=1}^{n} (x_r(i) - y(i))^2}}{\max(x_r) - \min(x_r)} \right) \times 100\% \tag{3.31}$$

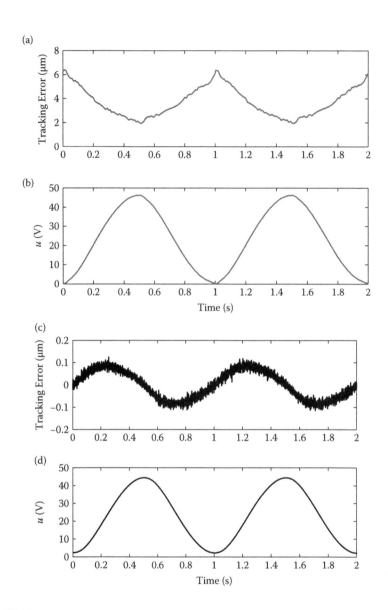

FIGURE 3.9
Experimental results of the feedforward controller and the PID controller. (a) Tracking error of the inversion feedforward controller. (b) Feedforward control signal. (c) Tracking error of the PID controller. (d) PID feedback control signal.

where n is the number of samplings, and $x_r(i)$ and $y(i)$ are the reference and the measured piezo displacement at the time instant i.

Figure 3.9(a, b) shows the tracking error and the feedforward control signal of the Preisach-based inversion feedforward controller. The RMS tracking error is 2.431 μm, the offset is 2.1 μm, and e_{rms} is 6.08%. It indicates that

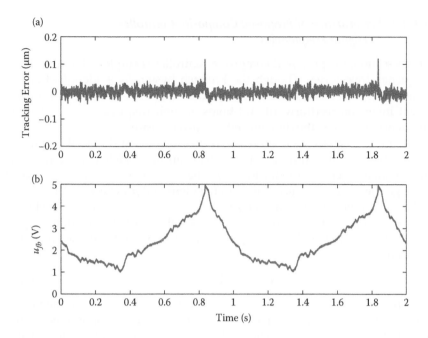

FIGURE 3.10
Experimental results of the proposed composite controller. (a) Tracking error. (b) Feedback control signal u_{fb} of the composite controller.

the inverse Preisach feedforward controller is effective to compensate the Preisach hysteresis, but the feedforward controller cannot reject disturbances and suppress modeling errors. Figure 3.9(c) shows the tracking performance of the PID controller in (3.30). The RMS tracking is 0.061 μm, and e_{rms} is 0.15%. The PID controller is effective to track low-frequency trajectories, but its performance is still limited by the Preisach hysteresis.

Figure 3.10 shows the tracking performance of the proposed composite controller. The RMS tracking error is reduced to 0.014 μm, the offset disturbance is also suppressed, and e_{rms} is 0.035%. The tracking error is reduced by 76.7%, compared to the PID controller in (3.30). The composite controller gives the best performance.

The feedback control signal u_{fb} of the composite controller, as shown in Figure 3.10(b), is less than 12% of the PID control signal in Figure 3.9(d). The feedforward control signals of the inversion feedforward controller and the composite controller are the same. The control signals in Figures 3.9(b, d) and 3.10(b) indicate that the Preisach-based feedforward eliminates the hysteresis and the PID feedback controller suppresses the disturbances and residual errors.

3.1.3.4 Performance of Proposed Composite Controller at Higher Frequencies

In this section, the proposed composite controller is employed to track high-frequency trajectories. Though the Preisach hysteresis is identified at low frequencies, the Preisach model in (2.1) is rate independent. Thus, the Preisach-based inversion feedforward still holds at high frequencies. As the input frequency increases, the dynamic effects also increase.

The piezo displacement is not the hysteresis output due to the dynamic effects, but the hysteresis in the PA also can be modeled using the rate-independent Preisach model in (2.1). The hysteresis is still compensated through the reference trajectories and the identified μ in Section 3.1.3.

The harmonic trajectory at 25 Hz is shown in Equation (3.32). According to the rate independence of Preisach hysteresis, the feedforward control signal of x_{r2} can be achieved by altering the timescale of the feedforward control signal of x_{r1}.

$$x_{r2}(t) = 20(1 + \sin 50\pi t) \tag{3.32}$$

At first, the same PID tuning controller in Equation (3.30) is applied to the piezoelectric stage. Figure 3.11(a, b) shows the tracking error and the control signal of the PID controller. The tracking error is significant, the RMS value is 1.187 μm, and e_{rms} is 3%.

Finally, the composite controller is implemented. Figure 3.11(c) shows the tracking error using the proposed composite controller, the RMS tracking error is 0.123 μm, and e_{rms} is 0.31%. e_{rms} is reduced by 89% compared to the PID controller. The composite controller gives the best tracking performance among the Preisach-based inversion feedforward controller, the PID controller, and the composite controller.

Figure 3.12 shows the curves of the actual displacement versus the desired displacement under the inversion feedforward, the PID controller, and the composite controller, respectively. The proposed composite controller achieves the smallest hysteresis curve, meaning the piezo tracks the displacement with the smallest delay and error.

3.1.4 Discussions

To achieve accurate motion control of PA mechanisms, this chapter presents the identification and compensation of Preisach hysteresis. The SVD-based algorithm is able to deal with an ill-conditioned mathematical issue due to the large number of discretized parameters on hand, and it can thus achieve more accurate estimation, but it requires a higher level of computational infrastructure for an online implementation to be viable. SVD-based parameter updating is employed to revise least-squares estimation with higher computing efficiency, and this approach improves the estimation performance.

The experimental results validate the soundness of Preisach hysteresis identification using the Preisach-based inversion feedforward. With the identified

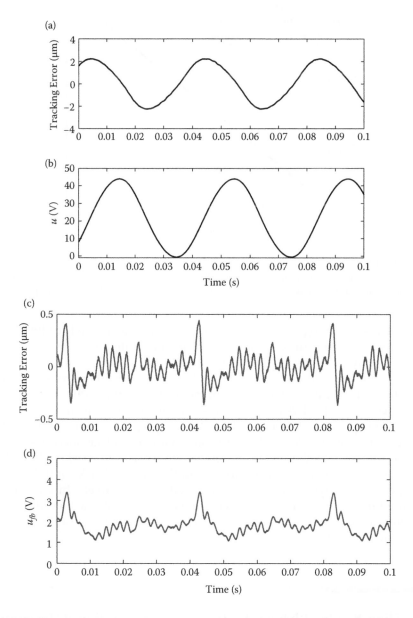

FIGURE 3.11
Experimental results of the PID controller and the composite controller. (a) Tracking error of the PID controller. (b) PID feedback control signal. (c) Tracking error of the composite controller. (d) Feedback control signal u_{fb} of the composite controller.

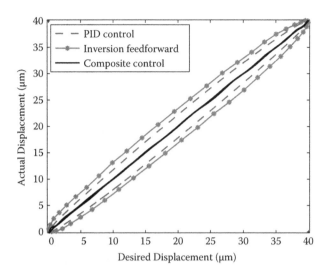

FIGURE 3.12
Actual versus desired displacement with the PID controller (dashed), the Preisach-based inversion feedforward controller (solid-dotted), and the composite controller (solid), respectively.

parameters and a log of the memory curve, a Preisach-based feedforward compensator is constructed that is complemented with a PID feedback controller. If only the inversion feedforward controller is used, the residual error and the dynamic effect due to vibration dynamics are not eliminated. If only the PID controller is used, the tracking performance of the PID controller is degraded due to the hysteresis effect. As the reference frequency increases, the tracking performance of the PID controller is degraded more significantly because of vibration effects. The composite controller still has adequate tracking performance as the reference frequency increases to 25 Hz, where the Preisach-based feedforward controller and the PID feedback controller can compensate the hysteresis and dynamics effect, respectively. In the experiment of the piezoelectric stage, accurate motion control is achieved by the proposed composite controller. The e_{rms} values at 1 Hz and 25 Hz are 0.035 and 0.31%, respectively.

3.2 High-Bandwidth Identification and Compensation of Hysteretic Dynamics in Piezoelectric Actuators

High-bandwidth and precision scanning are required in scanning probe microscopes (SPMs) for high-speed imaging and high-speed nanofabrication. However, the hysteretic dynamics in PAs greatly limits the scanning and tracking performance. In the open loop, the maximum error due to quasi-static hysteresis in PAs is 10–15% of the travel range [1]. As the input frequency increases, PAs exhibit dynamic response and the error from the hysteretic dynamics becomes more significant. Generally, a high-gain PID controller is adequate for scanning and tracking at low frequencies [29], but there is still a low-gain margin due to the rapid phase drop [30]. The closed-loop bandwidth attained using a simple feedback controller is typically less than 5–10% of the first resonant frequency because of complex hysteretic dynamics in PAs [31–33].

At high frequencies relative to the resonant frequencies, PAs exhibit dynamic and complex hysteresis. Thus, the rate-independent hysteresis model should be replaced with a dynamic type. Reference [34] presents a dynamic Preisach model to describe hysteresis at broadband frequencies, but the parameter identification process is complex and difficult. References [35] and [36] expand the rate-independent Prandtl–Ishlinskii hysteresis to a rate-dependent type, and density functions are assumed for parameter identification. Instead of expanding rate-independent hysteresis, the cascade connection of the rate-independent hysteresis and nonhysteretic dynamics is another approach to represent dynamic hysteresis over a broad range of frequencies [16]. We employ the cascaded model comprising the rate-independent Preisach hysteresis, electric and vibration dynamics to represent the hysteretic dynamics of PAs at broadband frequencies.

To push the application frontier of PAs toward fine scanning, various feedback control schemes have been investigated [37]. For instance, [38] and [39] employ a bounded uncertainty to represent the PA hysteresis, and propose a sliding mode controller and a robust adaptive controller to enhance tracking performance. Reference [40] presents the Prandtl–Ishlinskii hysteresis-based sliding model controller. However, fine and high-speed scanning is not simultaneously achieved with such approaches, since the hysteretic dynamics is not adequately compensated over a broad range of frequencies. Furthermore, feedback controllers have limited bandwidth. Alternatively, a model-based inversion feedforward controller can be employed to increase the scanning bandwidth and improve the scanning performance simultaneously, which relies on accurate model identification.

To achieve scanning at a rate higher than the first resonant frequency, this chapter proposes a composite controller consisting of a model-based inversion feedforward controller and a proportional–integral (PI) feedback controller. To design the composite controller, the PA model is identified first. The quasi-static hysteresis is identified using harmonic signals with varying amplitudes. The persistently exciting (PE) condition is satisfied with the harmonic input signals. Following this, the nonhysteretic dynamics is identified using a multifrequency harmonic input. Then, the composite controller is constructed based on the identified model. The inversion feedforward controller strictly depends on the hysteretic model and can be computed offline. The feedforward controller effectively expands the scanning bandwidth. To reject disturbances, the PI feedback controller is employed. The proposed composite controller presents simultaneous high-speed and precision scanning of PAs.

3.2.1 Proposed Model Identification Strategy

3.2.1.1 Model of PA Systems

The cascade connection is employed to represent the hysteretic dynamics in PA systems. Figure 3.13 illustrates the cascade structure. Γ, G_e, and G_v denote the quasi-static hysteresis, electric dynamics, and vibration dynamics, respectively. The nonhysteretic dynamics consists of G_e and G_v, which are assumed to be time invariant. u, v, and y represent the input voltage, hysteresis output, and piezo displacement, respectively. Furthermore, the hysteresis output v is unmeasurable. The quasi-static hysteresis is rewritten as the following rate-independent Preisach model:

$$v(t) = \iint_S \mu(\alpha, \beta) \gamma_{\alpha\beta}[u(t)] d\alpha d\beta \tag{3.33}$$

where S is the Preisach area, $\mu(\alpha, \beta)$ is the density function, and $\gamma_{\alpha\beta}$ is the hysteron output at point (α, β).

In this chapter, both the electric dynamics G_e and the vibration dynamics G_v are considered. Thus, the nonhysteretic dynamics $G = G_e G_v$ can be written as

$$G = \frac{k_{ev}}{\tau s + 1} \cdot \frac{\prod_j^{n-2} \left(s^2 + 2\xi_j \omega_j s + \omega_j^2\right)}{\prod_i^n \left(s^2 + 2\xi_i \omega_i s + \omega_i^2\right)} \tag{3.34}$$

where τ is the time constant of electric dynamics, ω_i and ω_j are the mode frequencies of vibration dynamics, ξ_i and ξ_j are the damping ratios, and k_{ev} is the DC gain of the nonhysteretic dynamics G.

The dynamics of PAs is quasi-static and almost rate independent at low frequencies, since the vibration and other high-frequency dynamics approach their DC gains at low-frequencies [41]. Thus, the key idea behind the proposed approach is to identify the Preisach hysteresis using low-frequency harmonic signals. Then, the hysteresis effects can be computed using the estimated Preisach model. Thereafter, the nonhysteretic dynamics is identified.

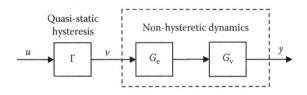

FIGURE 3.13
Block diagram of PA model.

3.2.1.2 Identification of Quasi-Static Hysteresis

The identification of a quasi-static hysteresis essentially entails the identification of the density function of the Preisach model (3.33). Based on the investigation in the previous section, the identification of a quasi-static hysteretic will be further studied in this section. The discretization method is also used for the ease of hysteresis identification. The hysteresis output in (3.33) at the time instant k can be represented as

$$v(k) = A_k X \qquad (3.35)$$

where $A_k = [\gamma_{11}(k), \gamma_{21}(k), \cdots, \gamma_{L1}(k), \cdots, \gamma_{LL}(k)]$, $X = [\mu_{11}s_{11}, \mu_{21}s_{21}, \cdots, \mu_{L1}s_{L1}, \cdots, \mu_{LL}s_{LL}]$, L is the discretization level of the Preisach plane, and γ_{ij}, μ_{ij}, and s_{ij} are the hysteron output, density value, and area of grid (i, j), respectively.

The identification of $\mu_{i,j}$ is equivalent to the estimation of $\mu_{ij}s_{ij}$, since s_{ij} is known. The data samples are collected at the time instants $t_1, t_2, \cdots, t_i, \cdots, t_N$, and a linear equation for Preisach hysteresis identification can be written as

$$AX = b \qquad (3.36)$$

where $A^{\mathrm{T}} = [A_1^{\mathrm{T}}, A_2^{\mathrm{T}}, \cdots, A_N^{\mathrm{T}}]$, $b^{\mathrm{T}} = [v(1), v(2), \cdots, v(N)]$.

The least-squares method is employed to estimate X. However, the PE problem is associated with a large number of parameters. This PE condition is mathematically equivalent to the singularity of $A^{\mathrm{T}}A$. Furthermore, the ill-condition due to small singular values of $A^{\mathrm{T}}A$ also has to be suppressed, and the SVD-based estimation is employed. The estimation of X in the least-squares sense is given by

$$\hat{X} = A^+ b \qquad (3.37)$$

where A^+ is the pseudoinverse of matrix A.

To perform the identification of quasi-static hysteresis, the low-frequency harmonic signal with varying amplitudes is proposed to eliminate the PE condition and solve equation (3.36) in the least-squares sense. Also, a sampling method that is uniform in the magnitudes of input voltages is also proposed.

For ease of identification, the variant input voltage is given in (3.38), whose amplitudes match the discretization of the Preisach plane.

$$u(t) = (P(t) - 0.5\delta) \frac{1 - \cos \omega_s t}{2} \qquad (3.38)$$

where $\delta = \max(u(t))/L$, ω_s is angular frequency, $(P(t) - 0.5\delta)$ is the amplitude at time t, and $P(t)$ is given by

$$P(t) = \left\lfloor 1 + \left(t - T_P \left\lfloor \frac{t}{T_P} \right\rfloor \right) \frac{\omega_s}{2\pi} \right\rfloor \delta \qquad (3.39)$$

where T_P is the period of $P(t)$, and $\lfloor \cdot \rfloor$ is the *floor* function that rounds the elements to the nearest integers in the direction of negative infinity.

The relationship between T_P and ω_s can be represented as

$$T_P = L\omega_s \tag{3.40}$$

The Preisach model (3.33) is not rate dependent but path dependent. Thus, the sampling for identification is not uniform with respect to (w.r.t.) time but uniform w.r.t. the input voltage $u(t)$. This sampling method is proven to be more effective in the experiment in Section 3.2.3.

The sampling instants of the ith harmonic signal are given by

$$t_{i,j} = \frac{1}{\omega_s} \arccos\left[1 - \frac{2(j - 0.5\delta)\delta}{P(t)} \right] + t_{i,1} \tag{3.41}$$

$$t_{i,2i-j} = \frac{1}{\omega_s} \left\{ 2\pi - \arccos\left[1 - \frac{2(j - 0.5\delta)\delta}{P(t) - 0.5} \right] \right\} + t_{i,1} \tag{3.42}$$

where $t_{i,1}$ is the starting time of the ith harmonic signal, $i = 1, 2, \cdots, P(t)/\delta$, $j = 1, 2, \cdots, i$, and $P(t) - 0.5\delta$ is the amplitude of the ith harmonic signal determined by Equation (3.39).

Adequate amplitudes are required to identify the Preisach hysteresis in Equation 3.35. Furthermore, the sampling is uniform w.r.t. the input signal due to the rate independence but path dependence of Preisach hysteresis.

The identification of the electric and vibration dynamics is performed at high frequencies. A multifrequency harmonic input is employed for adequate frequencies. The hysteresis output is estimated by the identified Preisach model. Figure 3.14 illustrates the estimation of the hysteresis output according to the identified Preisach model. $\hat{\Gamma}$ and $\hat{v}$ denote the identified Preisach model and the estimated hysteresis output, respectively. The identification of the electric and vibration dynamics is performed using $\hat{v}$ and y. Mature

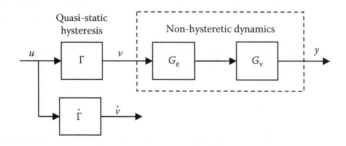

FIGURE 3.14
Output estimation of quasi-static hysteresis.

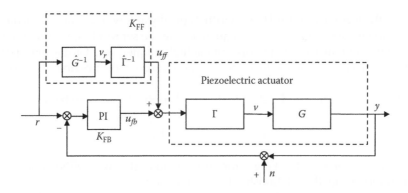

FIGURE 3.15
Composite control strategy.

identification methods can be employed in this part. For instance, the ARMAX method can be employed for parameter identification where the model structure is specified according to Equation (3.34).

3.2.2 Proposed Composite Controller

This section presents the design of the composite controller. The structure of the control system is shown in Figure 3.15; r denotes the reference trajectory, u_{ff} and u_{fb} denote the feedforward and feedback control signals, respectively, and $\hat{G}^{-1}$ and $\hat{\Gamma}^{-1}$ denote the estimated nonhysteresis dynamics and hysteresis nonlinearity, respectively. The composite controller consists of a primary model-based inversion feedforward controller and a secondary PI feedback controller. The feedforward controller is used to reduce phase lag and achieve high-speed scanning. It comprises the inversions of the nonhysteretic dynamics and the Preisach hysteresis. The feedforward controller is constructed based on the hysteresis and nonhysteretic dynamics. The discrete PI controller is employed for feedback that compensates disturbances and reduces modeling uncertainty within the feedback bandwidth.

3.2.2.1 Analysis of Feedforward and Feedback Controllers at High Frequencies

A model-based inversion feedforward controller is suitable to track high-frequency trajectories, because the feedforward controller is not affected by the measurement noise, which is outside of the feedforward control branch. In addition, the feedforward control signal approaches the inverse of the system gain. However, the feedforward controller is not capable of rejecting disturbances. Conversely, a feedback controller is capable of rejecting disturbances within the feedback control bandwidth, typically at low frequencies. Furthermore, the feedback gain at low frequencies can be set to a large value due to nonsignificant measurement noise, but at high frequencies, it is inappropriate to perform a large-gain feedback controller because of the possible chattering in case of noise.

The tracking error under the feedforward controller K_{FF} can be written as

$$\frac{e}{r}\Big|_{K_{FF}} = 1 - (G \circ \Gamma) K_{FF} \tag{3.43}$$

where $e = r - y$ is the tracking error, and $\circ$ denotes the composition operator, which represents the relationship between G and Γ.

Under the model-based inversion feedforward controller, the feedforward control signal u_{ff} can be represented as

$$u_{ff} = r(\hat{\Gamma}^{-1} \circ \hat{G}^{-1}) \tag{3.44}$$

where $\hat{G}^{-1}$ and $\hat{\Gamma}^{-1}$ denote the inversion of the estimation of the nonhysteretic dynamics and the Preisach hysteresis.

The feedforward control signal is not affected by measurement noise, and depends on the inversion gain of the PA model. With perfect inversion feedforward control, i.e., the inversion is sufficiently accurate, the feedforward controller can give adequate tracking without considering disturbances.

Alternatively, the tracking error under the feedback controller K_{FB} can be written as

$$\frac{e}{r}\Big|_{K_{FB}} = \frac{1}{1 + (G \circ \Gamma) K_{FB}} \tag{3.45}$$

The tracking error under the feedback controller K_{FB} can be reduced adequately if the control gain can be set to sufficiently large. At low frequencies, it is common to implement a feedback controller with a large integral action toward fine motion of PAs [37]. If the tracking error is suppressed by a factor of η, i.e., $e/r = 1/\eta$, the feedback control signal u_{fb} can be written as

$$u_{fb} = (n + r)(\eta - 1)(\Gamma^{-1} \circ G^{-1}) \tag{3.46}$$

where $\eta \gg 1$ for fine tracking, and n is broadband noise.

The feedback control signal u_{fb} in (3.46) indicates that high gains of the feedback controller at high frequencies easily result in instability due to the measurement noise n. Typically, the PA is a low-pass filter and the system gain is decreased at high frequencies, but the model-based inversion $\Gamma^{-1} \circ G^{-1}$ is a high-pass filter. Thus, if Equation (3.46) has $\eta \gg 1$ at high frequencies, the feedback control signal easily approaches chattering and saturation when the amplified high-frequency measurement noise coincides with the vibration modes of PAs.

Thus, we employ both the feedforward controller and the feedback controller to achieve high-speed and precision scanning. The feedforward controller will expand the scanning bandwidth, and the feedback controller will reject disturbances.

Remark: Model-based inversion feedforward controllers are suitable to expand tracking bandwidth and achieve high-speed scanning. Conversely, feedback controllers are suitable to reject unknown disturbances within the feedback bandwidth, and achieve precision scanning.

3.2.2.2 Design of Feedforward Controller

In this section, the model-based inversion feedforward controller is constructed to achieve high-speed scanning. It encompasses the inverse nonhysteretic dynamics and the inverse hysteresis. The reference signals pass through the inverse nonhysteresis dynamics $\hat{G}^{-1}$, then the inverse hysteresis $\hat{\Gamma}^{-1}$. The inversion of the nonhysteretic dynamics has more zeros than poles, and thus cannot be directly implemented for unknown and general reference signals in dSPACE board. To solve this problem, known and sufficiently smooth trajectories are employed. Furthermore, $\hat{v}_r$ can be computed offline.

After obtaining the inversion of the nonhysteresis dynamics, its output v_r is regarded as the reference of hysteresis inversion $\hat{\Gamma}^{-1}$. The feedforward voltage u_{ff} is regulated such that the unmeasurable hysteresis output v tracks v_r. The regulation is achieved using the hysteresis inversion based on the identified Preisach model. If the estimated hysteresis output $\hat{v}$ tracks the reference within the error range e_r, the feedforward voltage u_{ff} remains the same. If the estimated hysteresis output $\hat{v}$ is less than the reference v_r and $|v_r - \hat{v}| > e_r$, the feedforward voltage u_{ff} is increased and vice versa. Algorithm 3.1 illustrates the hysteresis inversion. The unmeasurable hysteresis output $\hat{v}$ is estimated

---

Algorithm 3.1: Inversion of the Identified Preisach Hysteresis

---

if $v_r(k) < \hat{v}(k)$ **then**
 $u_{ff}(k+1) = u_{ff}(k) - i\delta_h,$ $i = 1, \cdots, m$
 $\hat{v}(k+1) = \hat{\Gamma}(u_{ff}(k+1))$
 if $|v_r(k) - \hat{v}(k+1)| \leq e_r$ **then**
  break
 end if
else
 if $v_r(k) > \hat{v}(k)$ **then**
  $u_{ff}(k+1) = u_{ff}(k) + i\delta_h, i = 1, \cdots, m$
  $\hat{v}(k+1) = \hat{\Gamma}(u_{ff}(k+1))$
  if $|v_r(k) - \hat{v}(k+1)| \leq e_r$ **then**
   break
  end if
 else
  $u_{ff}(k+1) = u_{ff}(k)$
 end if
end if

---

using the identified density function $\hat{\mu}$. Parameter m is the total number of iterations in each regulation.

The iteration step δ_h is represented as

$$\delta_h = \lambda u_{\max}/L' \tag{3.47}$$

where λ is a coefficient to regulate the step, $u_{\max}$ is the maximum input voltage used in the hysteresis identification, and L' is the new discretization level.

3.2.2.3 Design of Feedback Controller

Though robust control and high gain control have been commonly studied to compensate the hysteresis and vibration dynamics of PA, they are complex and are not suitable to track trajectories faster than the resonant frequencies. The high-speed scanning performance will be achieved primarily by the model-based inversion feedforward controller. However, a feedback controller is also necessary to maintain stability and robustness in the face of disturbances and modeling errors. For disturbance rejection, a simple PI controller is designed according to the identified hysteretic dynamics. The more complex computations required for the hysteretic dynamics compensation are left outside of the feedback loop.

A traditional tuning method has been used to obtain a PID controller for feedback control. In this chapter, the PID tuning considering hysteretic dynamics will be investigated. Additionally, to reduce the sensitivity to high-frequency measurement noise, the derivative action is not used. Alternatively, the compensation performance at high frequencies is achieved by using the model-based inversion feedforward control.

The identified model of PAs is used to compute the ultimate gain and the ultimate period. Ziegler–Nichols (Z–N) tuning rules can be used to obtain the parameters of the PI controller. First, the ultimate gain and period are computed using the identified nonhysteretic dynamics. Then, the ultimate gain is adjusted according to the identified Preisach model, while the ultimate period is kept unchanged as it is not affected by the Preisach hysteresis.

3.2.2.3.1 Characteristic of PID Tuning in PAs

PAs have both rate-independent hysteresis and nonhysteresis dynamics. The hysteresis is represented by the Preisach model (3.33), which is static and path dependent. The static hysteresis does not exhibit dynamic responses, but will alter the input gain. Conversely, the electric and vibration dynamics exhibit dynamic responses. Thus, the PID tuning methods using step response or relay tuning will result in the uncertainty of ultimate gain in PAs. The gain uncertainty w.r.t. input voltage $u(t)$ due to Preisach hysteresis is illustrated in Figure 3.16. $S^{\max}$ is the activated area corresponding to the maximum gain for $u(t)$. Conversely, $S^{\min}$ is the area corresponding to the minimum gain for $u(t)$.

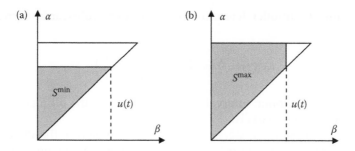

FIGURE 3.16
Illustration of the uncertain gain due to Preisach hysteresis. (a) Minimum gain. (b) Maximum gain.

The minimum and maximum gains of the input voltage $u(t)$ due to Preisach hysteresis are represented as

$$\begin{cases} \Delta_{max} = \iint\limits_{S^{max}} \mu(\alpha, \beta)\gamma_{\alpha\beta}[u(t)]d\alpha d\beta/u(t) \\ \Delta_{min} = \iint\limits_{S^{min}} \mu(\alpha, \beta)\gamma_{\alpha\beta}[u(t)]d\alpha d\beta/u(t) \end{cases} \tag{3.48}$$

where Δ_{min} and Δ_{max} denote the minimum and maximum gains due to the Preisach hysteresis.

Thus, the gain uncertainty Δ_h due to the Preisach hysteresis can be represented as

$$\Delta_{min} \leq \Delta_h \leq \Delta_{max} \tag{3.49}$$

3.2.2.3.2 Ultimate Gain and Period

At first, the ultimate period p_u and the gain margin g_m are determined at the cross-frequency ω_c of the nonhysteretic dynamics G:

$$p_u = \frac{1}{\omega_c} \tag{3.50}$$

where ω_c is the cross-frequency on the order of hertz.

The gain margin g_m of the nonhysteretic dynamics at the cross-frequency ω_c is given by

$$g_m = \frac{1}{|G(j\omega_c)|}$$

where $G(j\omega_c)$ is the gain of the nonhysteretic dynamics at the cross-frequency ω_c.

Then, the overall ultimate gain k_u is obtained through the gain margin g_m and the uncertainty Δ_{max}.

$$k_u = \frac{g_m}{\Delta_{max}} \tag{3.51}$$

Finally, the PI controller determined by Z–N tuning rules can be represented as

$$K_{PI}(z) = 0.6k_u \left(1 + \frac{2}{p_u} \frac{T}{2} \frac{z+1}{z-1}\right) \tag{3.52}$$

where T is the sampling interval, and k_u and p_u are the ultimate gain and ultimate period, respectively.

Remark: The hysteresis effect alters the ultimate gain of the PI controller determined by Z–N tuning rules. The ultimate gain can be modified by the identified Preisach hysteresis.

3.2.3 Experimental Studies

3.2.3.1 Identification of Quasi-Static Hysteresis

The piezoelectric hysteresis is quasi-static at low frequencies, and low-frequency harmonic signals are used to identify the rate-independent Preisach hysteresis. In Equations (3.38)–(3.42), $u_{max} = 60$ V, $L = 60$, $\delta = 1$, $\omega_s = 0.6$ Hz, and $T_P = 100$ s are used for the input voltage and the sampling in hysteresis identification.

3.2.3.2 Drift Suppression

There exists displacement drift in the piezoelectric stage. Figure 3.17(a) shows the drift effect in the piezo displacement untill 20 s. In this experiment, 100 sampling points are used and the sampling time is the starting time $t_{i,1}$ of the ith harmonic signal. The polynomial curve fitting is employed to suppress the drift. MATLAB function `polyfit` is used, and the polynomial for the drift suppression is given by

$$y_d(t) = 0.000121t^2 + 0.0127t - 0.013 \tag{3.53}$$

where $y_d(t)$ is the drift displacement, and t is the time. Figure 3.17(b) shows the drift suppression performance using the curve-fitting approach. The curve fitting gives satisfactory accuracy.

3.2.3.3 Preisach Hysteresis Identification

The discretization level L of the Preisach plane is set to 60. Then, the dimension of $A^T A$ is 1830×1830. In each grid (i, j), the density at the central point is regarded as the density of grid (i, j). To reduce the computing time, only the points determined by (3.41) and (3.42) are used. Extremes of the input signal are still preserved, thus preserving the hysteresis memory.

In hysteresis identification, only 3661 points are sampled. The rank of $A^T A$ is 1830. Thus, the PE condition is eliminated. Figure 3.18(a) shows the singular values of $A^T A$. The condition number of $A^T A$ is 1.92×10^7, which indicates $A^T A$ is ill-conditioned and small singular values should be truncated.

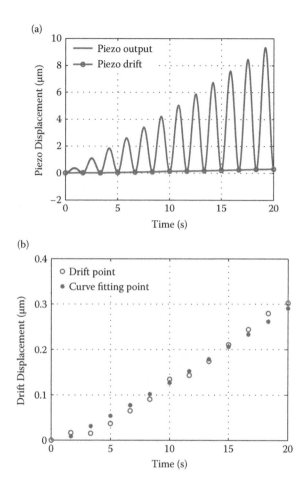

FIGURE 3.17
(a) Displacement drift. (b) Drift suppression.

The singular value selection is a trade-off problem. A larger truncation threshold increases the matrix approximation error but suppresses the ill-conditioned problem. Conversely, a smaller truncation threshold decreases the matrix approximation error but results in a more serious ill-conditioned problem. In this chapter, the truncation is implemented according to the error analysis of hysteresis identification. The relative error between the estimated displacement error and the measured displacement is employed to describe the hysteresis identification accuracy, because the density function μ is unknown.

The relative error is given by

$$e_h(\%) = \frac{\|y - \hat{y}\|}{\|y\|} \tag{3.54}$$

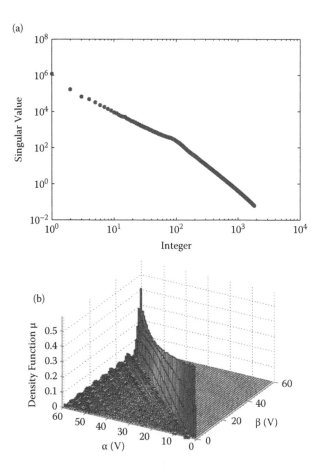

FIGURE 3.18
(a) Singular values of $A^T A$. (b) The identified density function.

where y and $\hat{y}$ are the measured and estimated displacements. With trials, the singular values smaller than 1.2 are truncated to suppress the ill-conditioned problem. Equation (3.37) is employed to identify the density function. Figure 3.18(b) shows the identified density function. The relative error in (3.54) is less than 3%, and the Preisach hysteresis identification is satisfactory.

3.2.3.4 Identification of Nonhysteresic Dynamics

The nonhysteretic dynamics is identified using a multifrequency input. The output $\hat{v}$ of the quasi-static hysteresis is computed using the identified Preisach model. First, to choose the input frequency, a relay feedback is used to estimate the ultimate frequency of the piezoelectric stage. Figure 3.19 shows the sketch of relay feedback where the relay output switches between 0 and 10 V. From the experiment, the estimated ultimate frequency is 500 Hz. The

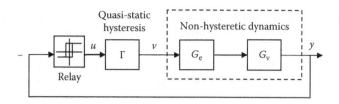

FIGURE 3.19
Ultimate frequency estimation using relay feedback.

harmonic signal with adequate frequencies is required to excite more modes. In the experiment, two vibration modes are identified. According to the estimated ultimate frequency, the multifrequency input signal consisting of six sinusoids is given by

$$u_d(t) = a_0 + \sum_{i=1}^{6} a_i \sin 2\pi f_i t$$

where $a_0 = 24$, $a_1 = 3$, $a_2 = 6$, $a_3 \sim a_6 = 3$, $f_1 = 100$ Hz, $f_2 = 200$ Hz, $f_3 = 300$ Hz, $f_4 = 400$ Hz, $f_5 = 500$ Hz, and $f_6 = 600$ Hz.

To estimate the hysteresis output accurately, the Preisach plane is rediscretized to be 180×180, i.e., $L' = 180$. The estimated hysteresis output $\hat{v}$ based on the identified Preisach model is regarded as the input to the nonhysteretic dynamics.

The nonhysteretic dynamics is identified using the ARMAX method [42], where the denominator, the numerator, and the error are identified with orders of 5, 1, and 5, respectively. The estimated parameters of the nonhysteretic dynamics are $\hat{k}_{ev} = 0.8716$, $\hat{\tau} = 0.000474$ s, $\hat{\omega}_1 = 453.5$ Hz, $\hat{\omega}_2 = 792.7$ Hz, $\hat{\xi}_1 = 0.67$, and $\hat{\xi}_2 = 0.081$, respectively. Figure 3.20 shows the comparison

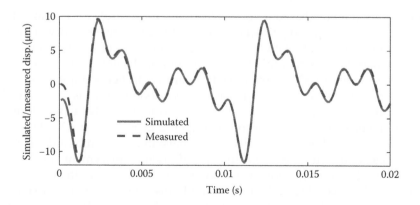

FIGURE 3.20
Comparison of simulated and measured displacement.

of the simulated and measured displacements with zero mean value. It can be seen that the simulated displacement accurately matches the measured displacement.

Remark: The detailed discretization gives more smooth and precise estimation of the hysteresis output. In the experiments, the discretization level is 60 for the hysteresis identification, but the discretization level is 180 for the hysteresis estimation.

3.2.3.5 Controller Design

For easy application, harmonic signals are used as the scanning trajectories. The proposed composite controller is designed and implemented in the dSPACE board 1104. The composite controller consists of a mode-based inversion feedforward controller and a PI feedback controller. Based on the identified Preisach hysteresis in Section 3.2.3 and the nonhysteretic dynamics in Section 3.2.3, the mode-based inversion feedforward is constructed; λ, u_{max}, and L' are set to 1, 60, and 180 in Algorithm 3.1, respectively.

In order to suppress disturbances, the simple PI feedback controller in Section 3.2.2 is augmented. Based on the identified nonhysteretic model of the piezoelectric stage, the ultimate period is 0.0018 s. The maximum gain Δ_{max} due to Preisach hysteresis is 1.06. The ultimate gain k_u is 1.65. Finally, the following discrete PI controller is given by

$$K_{PI}(z) = \frac{1.01z - 0.976}{z - 1} \tag{3.55}$$

where the sampling interval is 0.025 ms.

3.2.3.6 Performance Evaluation

3.2.3.6.1 Proposed Inversion Feedforward Controller

The proposed model-based inversion feedforward controller is employed to reduce the phase lag and gain distortion. Furthermore, the feedforward controller can also be employed to validate the soundness of the model identification. Based on the identified hysteresis and nonhysteretic dynamics, the proposed model-based inversion feedforward controller is implemented in the piezoelectric stage. For comparison, open-loop control is also considered, and it only employs a positive value to regulate the actual desired displacement gain to one. The positive gain 2.22 is used at 600 Hz, such that the actual and desired displacements have the same amplitude. Figure 3.21(a) shows the curve of the actual and desired displacements under the model-based inversion feedforward controller. Compared to the open-loop response, the phase-lag and the magnitude distortion are reduced significantly by the model-based inversion feedforward controller. The effectiveness

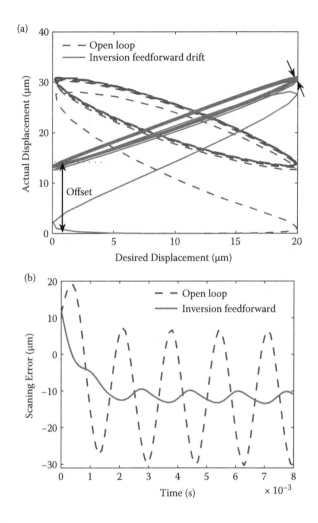

FIGURE 3.21
Performance of the model-based inversion feedforward controller.

of the model-based inversion feedforward controller also validates the soundness of the model identification. Figure 3.21(b) shows the scanning error at 600 Hz. The root mean square (RMS) scanning error of the open-loop control is 18.7 μm. The RMS scanning error of the model-based inversion feedforward is 13.54 μm, and it is reduced by 27.5%. Figure 3.21(a) also shows the offset and drift under the inversion feedforward controller. The offset is 13.5 μm and the maximum drift range is 1.1 μm, which will be suppressed using the PI feedback controller.

3.2.3.6.2 *Proposed Composite Controller*

The proposed composite controller is implemented. Moreover, the RMS error e_{rms} as a percentage of the output range is defined as

$$e_{rms}(\%) = \left(\frac{\sqrt{\frac{1}{p} \sum_{i=1}^{p} (y_r(i) - y(i))^2}}{\max{(y_r)} - \min{(y_r)}} \right) \times 100\% \tag{3.56}$$

where p is the number of sampling points, and $y_r(i)$ and $y(i)$ are the desired and actual displacements at time instant i, respectively. Figure 3.22(a–c) shows the curves of the actual versus the desired displacements with the above three different controllers at 40, 100, and 600 Hz, respectively. Figure 3.22(d–f) and Table 3.1 show the corresponding scanning performances at 40, 100, and 600 Hz, respectively. The proposed composite controller achieves both high-speed and precision scanning. Compared with the open-loop case, the PI controller improves the scanning performance, but the phase lag and scanning error of the PI controller also increase significantly as the scanning frequency increases. Thus, the PI controller has limited scanning bandwidth and performance at frequencies higher than the first resonant frequency. The composite controller presents the best performance. Both the phase lag and the scanning error are reduced significantly. Compared with the PI controller, the scanning errors are reduced by 88.1, 75.5, and 79.7% at frequencies 40, 100,

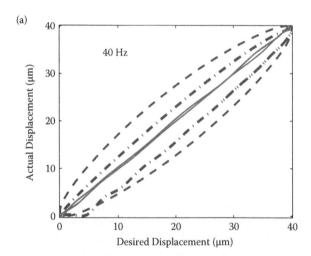

FIGURE 3.22
Scanning performance at different frequencies under open-loop control (dashed), PI control (dash-dotted), and composite control (solid). (a–c) Curves of the actual displacement versus the desired displacement. (d–f) Scanning error. (*continued*)

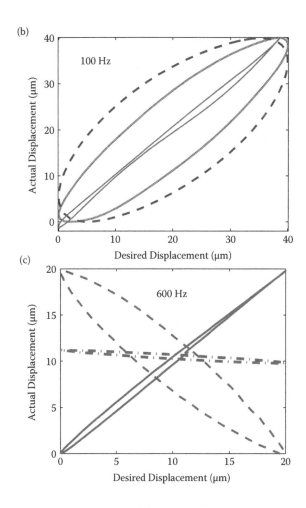

FIGURE 3.22
(*continued*). Scanning performance at different frequencies under open-loop control (dashed), PI control (dash-dotted), and composite control (solid). (a–c) Curves of the actual displacement versus the desired displacement. (d–f) Scanning error. (*continued*)

and 600 Hz, respectively. Specially, fine scanning performance is achieved by the composite controller. The RMS scanning error at 600 Hz is 0.352 μm and e_{rms} is 1.76%.

Remark: The model-based inversion feedforward controller is successful in expanding the scanning bandwidth higher than the resonant frequency of the PA. Furthermore, the PI feedback controller is successful in rejecting known disturbances within the feedback bandwidth.

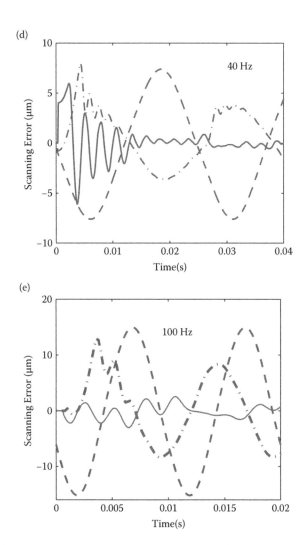

FIGURE 3.22

(*continued*). Scanning performance at different frequencies under open-loop control (dashed), PI control (dash-dotted), and composite control (solid). (a–c) Curves of the actual displacement versus the desired displacement. (d–f) Scanning error. (*continued*)

3.2.4 Discussions

The experimental results confirm that the proposed composite controller provides both high-speed and precision scanning at a rate higher than the resonant frequency. By applying the proposed composite controller, the scanning error e_{rms} is reduced to 0.30, 1.21, and 1.76% of the desired amplitudes at the

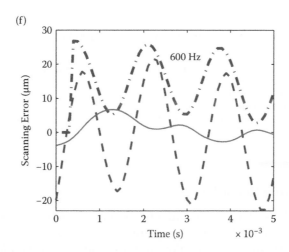

FIGURE 3.22

(*continued*). Scanning performance at different frequencies under open-loop control (dashed), PI control (dash-dotted), and composite control (solid). (a–c) Curves of the actual displacement versus the desired displacement. (d–f) Scanning error.

scanning frequencies 40, 100, and 600 Hz (8.84, 22.1, and 132.0% of the first resonant frequency), respectively.

Currently, the scanning frequency with satisfactory performance in most works is lower than the resonant frequency. For instance, Wu and Zou presented the robust inversion-based 2-DOF (degrees of freedom) control for the PA with a resonant frequency of 800 Hz [41], but the electric dynamics and hysteresis are not modeled. The scanning error e_{rms} for the small range (5 μm) trajectory at 300 Hz is 5.28%, as shown in Table 3.2. Results at higher scanning frequencies are not provided in the PA experiment [41]. Leang and Devasia proposed the high-gain feedback and inverse feedforward control for the atomic force microscopy (AFM) piezoelectric actuator with a resonant frequency of 486 Hz [2], as shown in Table 3.2; the scanning error e_{rms} at 450 Hz is 10.15%, while results at higher scanning frequencies are still not provided.

TABLE 3.1

Scanning Errors RMS (μm) and e_{rms} (%)

Frequency	Open-Loop	PI Control	Composite Control
40 Hz	5.144 (12.9%)	2.526 (6.31%)	0.30 (0.75%)
100 Hz	11.616 (29.1%)	6.218 (15.6%)	0.484 (1.21%)
600 Hz	14.961 (74.8 %)	8.014 (40.1%)	0.352 (1.76%)

TABLE 3.2

Comparison of Scanning Frequency and Relative Scanning
Error (%)

	Ref. [41]	Ref. [2]	Proposed Method
Resonant frequency	800 Hz	486 Hz	453.5 Hz
Scanning frequency	300 Hz	450 Hz	600 Hz
Relative RMS error (%)	5.28%	10.25%	1.76%

3.3 Concluding Remarks

Based on the Preisach model, the identification and compensation of the hysteretic phenomenon at low frequency are first addressed in this chapter. The SVD-based least-squares estimation and revision are adopted for the identification of these parameters. The Preisach-based inversion feedforward is developed to compensate the hysteresis in PAs. Additionally, a PID feedback controller is also augmented to suppress residual errors, disturbances, and dynamics effects. Experimental studies have been done to highlight the relative strengths of these algorithms with a view toward real-time precise control tracking applications.

The second part in this chapter presents an approach amenable to practical applications for dynamic hysteresis identification and high-speed motion control at higher frequencies. The feedforward controller, employing a model-based inversion, greatly extends the control bandwidth, while the PI feedback controller suppresses disturbances. The identification methodology and the composite control scheme are fully implemented on a real piezoelectric stage using a dSPACE control platform. The proposed composite controller achieves precision scanning at a rate higher than the resonant frequency.

References

1. Y. Li and Q. Xu. Adaptive sliding mode control with perturbation estimation and PID sliding surface for motion tracking of a piezo-driven micromanipulator. *IEEE Transactions Control Systems Technology*, 18(4), 798–810, 2010.
2. K. K. Leang and S. Devasia. Feedback linearized inverse feedforward for creep, hysteresis and vibration compensation in AFM piezoactuators. *IEEE Transactions Control Systems Technology*, 15(5), 927–935, 2007.
3. M. Goldfarb and N. Celanovic. Modeling piezoelectric stack actuators for control of micromanipulation. *IEEE Control Systems Magazine*, 17(3), 69–79, 1997.
4. D. Davino, C. Natale, S. Pirozzi, and C. Visone. Phenomenological dynamic model of a magnetostrictive actuator. *Physics B: Condensed Matter*, 343(1), 112–116, 2004.

5. A. Visintin. *Differential models of hysteresis*. New York: Springer-Verlag, 1996.

6. I. Mayergoyz. *Mathematical models of hysteresis*. Berlin: Springer-Verlag, 2003.

7. H. Liaw and B. Shirinzadeh. Neural network motion tracking control of piezo actuated flexure based mechanisms for micro nanomanipulation. *IEEE/ASME Transactions on Mechatronics*, 14(5), 517–527, 2009.

8. S. Mittal and C. Menq. Hysteresis compensation in electromagnetic actuators through Preisach model inversion. *IEEE/ASME Transactions on Mechatronics*, 5(4), 394–408, 2000.

9. K. Kuhnen. Modeling, identification and compensation of complex hysteretic nonlinearities: A modified Prandtl-Ishlinskii approach. *European Journal of Control*, 9(4), 407–418, 2003.

10. C. Visone. Hysteresis modeling and compensation for smart sensors and actuators. *Journal of Physics: Conference Series*, 138(1), 012028, 2008.

11. A. Putra, S. Huang, K. K. Tan, and T. H. Lee. Design, modeling, and control of piezoelectric actuators for intracytoplasmic sperm injection. *IEEE Transactions Control Systems Technology*, 15(5), 879–890, 2007.

12. F. Ikhouane and J. Rodellar. A linear controller for hysteresis systems. *IEEE Transactions on Automatic Control*, 51(2), 340–344, 2006.

13. H. Hu. Compensation of hysteresis in piezoceramic actuators and control of nanopositioning system. PhD dissertation, University of Toronto, Toronto, 2003.

14. G. Song, J. Zhao, and X. Zhou. Tracking control of a piezoceramic actuator with hysteresis compensation using inverse Preisach model. *IEEE/ASME Transactions on Mechatronics*, 10(2), 198–209, 2005.

15. R. Iyer, X. Tan, and P. Krishnaprasad. Approximate inversion of Preisach hysteresis operator with application to control of smart actuators. *IEEE Transactions on Automatic Control*, 50(6), 798–809, 2005.

16. X. Tan and J. S. Baras. Adaptive identification and control of hysteresis in smart materials. *IEEE Transactions on Automatic Control*, 50(6), 827–839, 2005.

17. O. Henze and W. M. Rucker. Identification procedures of Preisach model. *IEEE Transactions on Magnetics*, 38(2), 833–836, 2002.

18. K. Astrom and B. Wittenmark. *Adaptive control*. New York: Addison-Wesley, 1994.

19. R. Aster, B. Borcher, and C. Thurber. *Function analysis and inverse problem*. Amsterdam: Elsevier, 2005.

20. S. S. Aphale, B. Bhikkaji, and S. R. Moheimani. Minimizing scanning errors in piezoelectric stack-actuated nanopositioning platforms. *IEEE Transactions on Nanotechnology*, 7(1), 79–90, 2008.

21. G. M. Clayton, S. Tien, K. K. Leang, Q. Zou, and S. Devasia. A review of feedforward control approaches in nanopositioning for high-speed SPM. *ASME Journal of Dynamic Systems Control*, 131, 0611011, 2009.

22. G. Tao and P. Kokotovic. Adaptive control of plants with unknown hysteresis. *IEEE Transactions on Automatic Control*, 40(2), 200–212, 1995.

23. X. Chen, T. Hisayama, and C. Su. Pseudo-inverse-based adaptive control for uncertain discrete time systems preceded by hysteresis. *Automatica*, 45, 469–476, 2009.

24. J. R. Bunch and C. P. Nielsen. Updating the singular value decomposition. *Numerical Mathematics*, 31(2), 111–129, 1978.

25. M. Brand. Fast online SVD revisions for lightweight recommender systems. In *SIAM International Conference on Data Mining*, 2003, pp. 37–48.

26. J. Lin, H. Chiang, and C. C. Lin. Tunnung PID control gains for micro piezo-stage using grey relational analysis. In *International Conference on Machine Learning and Cybernetics*, July 2008, pp. 3863–3868.
27. Y. Wu and Q. Zou. Iterative control approach to compensate for both the hysteresis and the dynamics effects of piezo actuators. *IEEE Transactions Control Systems Technology*, 15(5), 936–944, 2007.
28. L. Liu, K. K. Tan, S. N. Huang, and T. H. Lee. Identification and control of linear dynamics with input Preisach hysteresis. In *American Control Conference*, June 2010, pp. 4301–4306.
29. H. Xie, M. Rakotondrabe, and S. Regnier. Characterizing piezoscanner hysteresis and creep using optical levers and a reference nanopositioning stage. *Review of Scientific Instruments*, 80(4), 046102, 2009.
30. K. Leang, Q. Zou, and S. Devasia. Feedforward control of piezoactuators in atomic force microscope systems. *IEEE Control Systems Magazine*, 29(1), 70–82, 2009.
31. A. Fleming. Nanapositioning system with force feedback for high performance tracking and vibration control. *IEEE/ASME Transactions on Mechatronics*, 15(3), 433–446, 2010.
32. W. Kuo, S. Chuang, C. Nian, and Y. Tarng. Precision nano-alignment system using machine vision with motion controlled by piezoelectric motor. *Mechatronics*, 18(1), 21–34, 2008.
33. I. A. Mahmood, S. O. Reza Moheimani, and B. Bhikkaji. A new scanning method for fast atomic force microscopy. *IEEE Transactions on Nanotechnology*, 10(2), 203–216, 2011.
34. I. Mayergoyz. Dynamic Preisach models of hysteresis. *IEEE Transactions on Magnetics*, 24(6), 2925–2927, 1988.
35. M. Janaiden, C. Su, and S. Rakheja. Development of the rate-dependent Prandtl-Ishlinskii model for smart actuators. *Smart Materials and Structures*, 17, 035026, 2008.
36. M. A. Janaideh, S. Rakheja, and C. Y. Su. An analytical generalized Prandtl-Ishlinskii model inversion for hysteresis compensation in micropositioning control. *IEEE/ASME Transactions on Mechatronics*, 16(4), 734–744, 2011.
37. S. Devasia, E. Eleftheriou, and S. Moheimani. A survey of control issues in nanopositioning. *IEEE Transactions Control Systems Technology*, 15(5), 802–823, 2007.
38. H. Liaw and B. Shirinzadeh. Sliding mode control enhanced adaptive motion tracking control of piezoelectric actuation systems for micro/nano manipulation. *IEEE Transactions Control Systems Technology*, 16(4), 826–833, 2008.
39. H. C. Liaw and B. Shirinzadeh. Robust adaptive constrained motion tracking control of piezo-actuated flexure-based mechanisms for micro/nano manipulation. *IEEE Transactions on Industrial Electronics*, 58(4), 1406–1415, 2011.
40. X. Chen and T. Hisayama. Adaptive sliding-mode position control for piezo-actuated stage. *IEEE Transactions on Industrial Electronics*, 55(11), 3927–3934, 2008.
41. Y. Wu and Q. Zou. Robust inversion-based 2-DOF control design for output tracking: Piezoelectric-actuator example. *IEEE Transactions Control Systems Technology*, 17(5), 1069–1082, 2009.
42. L. Ljung. *System identification: Theory for the user*. 2nd ed. Upper Saddle River, NJ: Prentice-Hall, 1999.

4

Identification and Compensation of Friction and Ripple Force

Various nonlinear factors originate from the actuators of motion control systems, such as stiction, friction, hysteresis, and force ripple, which affect their performance. It is important to model these behaviors and use appropriate control schemes to eliminate these effects to increase the precision of the motion control systems.

The main objective of this chapter lies in the enhancement of the accuracy of motion control systems by proposing and identifying the models for commonly encountered nonlinearities, such as frictions and force ripples, efficiently and accurately, with relay feedback approaches.

4.1 Relay Feedback Techniques for Precision Motion Control

The relay feedback technique has been introduced in control application since the 1960s. Although the theoretical studies of relay feedback systems have been made with great leaps since the 1970s, the applications of relay feedback are mainly limited to design of adaptive controllers [1] and autotuning of proportional-integral-derivative (PID) controllers [2]. The principle behind relay-based PID autotuning is simple; self-oscillation is generated with relay elements, from which the system characteristics are inferred and subsequently used to tune the controller. The simplest form of relay feedback system is shown in Figure 4.1. The most important application of relay feedback system (RFS) is design of autotuners for PID controllers [2,3,5,16], where continuous cycling of the controlled variable is generated from the relay experiment and the important model information can be directly extracted from it. Compared with the conventional Ziegler–Nichols tuning, the sustained oscillation generated in the relay experiment is in a control manner and a very efficient way, i.e., a one-shot solution. The simplicity of the tuning mechanism makes the relay-based autotuner a great success. The various commonly used single relay elements are shown in Figure 4.2, including single-valued relay, hysteresis

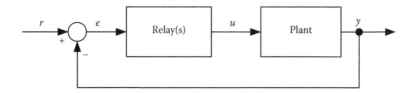

FIGURE 4.1
The simplest form of RFS.

relay, and dead-zone relay. Recent research tries to use relay feedback systems for modeling of nonlinear hybrid systems, typically friction-impeded motion control systems, by the same basic principles.

4.2 Identification and Compensation of Friction Model

The designs and applications of motion control systems have been closely related to investigation of friction between contact surfaces of a machine's subparts. Thus, this section will focus on this area.

4.2.1 System Model

The dynamics of a servomechanical system is described using a nonlinear mathematical model:

$$u(t) = K_e \dot{x} + Ri(t) + L di(t)/dt \qquad (4.1)$$

$$f(t) = K_f i(t) \qquad (4.2)$$

$$f(t) = m\ddot{x}(t) + \bar{f}_{load}(t) + \bar{f}_{nl}(t) \qquad (4.3)$$

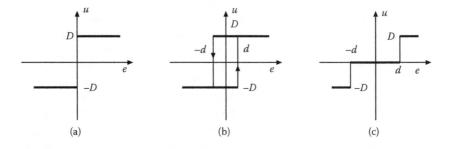

FIGURE 4.2
Variations of relay elements. (a) Relay without hysteresis. (b) Relay with hysteresis. (c) Relay with dead zone.

TABLE 4.1

Parameters of the Linear Motor

Parameters	Physical Meaning	Units
K_f	Force constant	N/amp
R	Resistance	ohms
K_e	Back EMF	vol/m/s
L	Armature inductance	mh
m	Mass of moving part	kg

where $u(t)$ and $i(t)$ are the time-varying motor terminal voltage and armature current, respectively; $x(t)$ is the motor position; $f(t)$ and $\bar{f}_{load}$ are the developed force and the applied load force, respectively, and $\bar{f}_{nl}$ is nonlinearity affecting the developed force. In the servomechanical system concerned in this chapter, friction force $\bar{f}_{fric}$ and the remaining small and unaccounted dynamics $\bar{f}_{res}$ are presented. Thus,

$$\bar{f}_{nl} = \bar{f}_{fric} + \bar{f}_{res}. \tag{4.4}$$

Other parameters are described in Table 4.1.

Since the electrical time constant is much smaller than the mechanical one, the transient delay due to the electrical response is ignored. The following equivalent model is obtained after simplification:

$$\ddot{x} = (a\dot{x} + u - f_{fric} - f_{load} - f_{res})/b \tag{4.5}$$

where $a = -K_e$, $b = mR/K_f$, $f_{fric} = R\bar{f}_{fric}/K_f$, $f_{load} = R\bar{f}_{load}/K_f$, and $f_{res} = R\bar{f}_{res}/K_f$.

Let $\tilde{u} = u - f_{fric} - f_{load} - f_{res}$, $\tau = -b/a$, and $K = -1/a$. The transfer function of the linear portion of the servomechanical system is shown to be

$$G_p(s) = X(s)/\tilde{U}(s) = K/[s(\tau s + 1)] \tag{4.6}$$

The friction force is usually modeled as an odd nonlinearity with different types of friction components. The complexity and required accuracy of the model mainly depend on the application domain. When the system operates essentially in the high-velocity mode, a two-parameter friction model, which takes into account the Coulomb friction f_c and viscous friction f_v [4,14], is adequate enough. However, when the system operates in the low-velocity or a bidirectional mode, a more accurate and elaborate model, which considers the static friction f_s, Coulomb friction f_c, viscous friction f_v, as well as the Stribeck effect, will become necessary [4,10].

The generalized friction force f, discussed in the method, is a summation of friction force f_{fric} and loading force f_{load}. If the loading force is dependent on the direction of motion, f_{load} is described as $f_{load} = f_l \mathrm{sgn}(\dot{x})$. The generalized

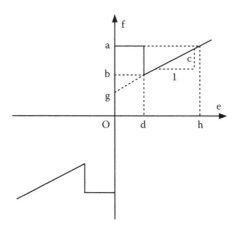

FIGURE 4.3
Four-parameter friction model.

four-parameter friction model as shown in Figure 4.3 is expressed as

$$f = \begin{cases} f_1 \operatorname{sgn}(\dot{x}) & \text{if } |\dot{x}| < \delta \\ [f_2 + f_3(|\dot{x}| - \delta)] \operatorname{sgn}(\dot{x}) & \text{if } |\dot{x}| \geq \delta \end{cases} \tag{4.7}$$

where f_1 is the generalized maximum static friction, f_2 is the generalized Coulomb friction, f_3 is associated with the viscosity constants, and δ is the lubrication boundary velocity (LBV), where $f_1 = f_s + f_l, f_2 = f_c + f_l, f_3 = f_v$.

4.2.2 DCR Feedback System

The dual-channel relay (DCR) feedback structure, as shown in Figure 4.4, has been used for the identification of the friction model as well as the parameters of the linear dynamical part of a servomechanical system [8,14].

For the convenience of further discussion, an equivalent circuit is shown in Figure 4.4(b) that segregates the full feedback system into a linear portion and a nonlinear portion. The linear portion contains the system dynamics and DC gain, while the nonlinear portion includes the actual frictional and load forces system relay (SR), as well as the two intentional relays feedback relay (FR) FR1 and FR2 in use. The describing function (DF) of the equivalent relay (N_{ER}) is simply the sum of the individual DFs due to the feedback relays (N_{FR1}, N_{FR2}) and the inherent system relay (N_{SR}), i.e.,

$$N_{ER} = N_{FR1} + N_{FR2} + N_{SR} \tag{4.8}$$

where $N_{FR1}(A) = 4h_1/(\pi A)$, $N_{FR2}(A) = -4jh_2/(\pi A)$. Similar to [10], the nonlinear friction element in the four-parameter friction model of Figure 4.3 is approximated with quasi-linear elements by using the following DFs [9,12],

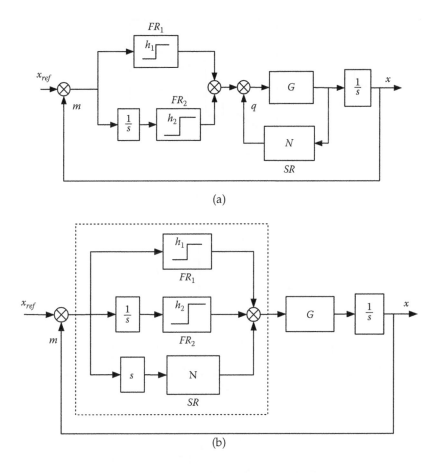

FIGURE 4.4
The DCR apparatus. (a) Original setup. (b) Equivalent system.

as shown in Figure 4.5:

$$N_{SR}(A, \omega) = N_A(A) + N_B(A, \omega) - N_C(A, \omega) \quad (4.9)$$

where

$$N_A(A) = 4jf_1/(\pi A) \quad (4.10)$$

$$N_B(A, \omega) = \begin{cases} 0 & , \omega A < \delta \\ \frac{2jf_3\omega}{\pi} \left[\cos^{-1}\left(\frac{\delta}{\omega A}\right) - \frac{\delta}{\omega A}\sqrt{1 - \left(\frac{\delta}{\omega A}\right)^2} \right] & , \omega A \geq \delta \end{cases} \quad (4.11)$$

$$N_C(A, \omega) = \begin{cases} 0 & , \omega A < \delta \\ \frac{4j(f_1 - f_2)}{\pi A}\sqrt{1 - \left(\frac{\delta}{\omega A}\right)^2} & , \omega A \geq \delta \end{cases} \quad (4.12)$$

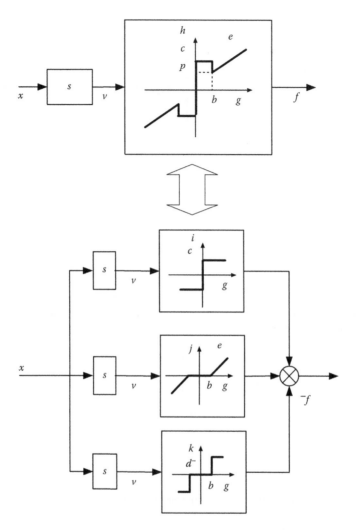

FIGURE 4.5
Friction model decomposition.

Remark 1: The DFs (N_B and N_C) are frequency dependent, compared to those in [10], since the inherent relay due to friction is precascaded with a differentiator. This arises because position feedback is used in the method, instead of the velocity feedback used in [10].

Remark 2: The DFs (N_B and N_C) are piecewise continuous, and ωA is an approximation of the velocity amplitude. This is reasonable since the DF analysis assumes a sinusoidal input $x(t) = A\sin(\omega t)$, which, after differentiation and before input to the relay element, becomes $\dot{x}(t) = \omega A\cos(\omega t)$.

4.2.3 Friction Modeling Using DCR Feedback

The proposed procedure to yield the system model from relay feedback experiments comprises two phases.

4.2.3.1 Low-Velocity Mode: Static Friction Identification

As discussed in the previous section, if the amplitude of velocity is kept below δ (i.e., $\omega A < \delta$ holds), the DF of the friction model can be simplified as $N_{SR}(A) = 4jf_1/(\pi A)$, which is frequency independent. Thus, it is possible to estimate the generalized static friction model if $\omega A < \delta$.

In order to identify K, τ, and f_1, two relay experiments are to be conducted with the system operating in low-velocity mode with $\omega A < \delta$. The amplitudes and frequencies of limit cycles are A_1, A_2 and ω_1, ω_2 accordingly, and the gains of relays are represented as h_{11}, h_{21}, h_{12}, and h_{22} accordingly, where h_{ij} denotes the gain of the ith relay during the jth experiment. From the proposed procedures, it follows that

$$K = \pi(\omega_2 A_2 - \omega_1 A_1)/[4(h_{22} - h_{21})] \tag{4.13}$$

$$f_1 = (h_{21}\omega_2 A_2 - h_{22}\omega_1 A_1)/(\omega_2 A_2 - \omega_1 A_1) \tag{4.14}$$

Since there are four equations but only three unknowns, two equations are given to compute τ. An averaging approach is used, so that

$$\tau_l = 2K\left[h_{11}/\left(A_1\omega_1^2\right) + h_{12}/\left(A_2\omega_2^2\right)\right]/\pi \tag{4.15}$$

where τ_l is the time constant of the linear dynamics of the system estimated in this first phase, with the system operating in a low-velocity mode.

It is efficient to estimate the three parameters explicitly via (4.13)~(4.15). However, the estimation is based on the assumption that $\omega A < \delta$.

In summary, a systematic set of procedures to select appropriate relay gains in the low-velocity experiments, is prescribed as follows:

1. Select a small enough h_2 to ensure $A_v < \delta$.
2. Select a small h_1 to reduce ω and increase A while maintaining a small A_v and an adequate signal-to-noise ratio (SNR).

One may run this first phase of the relay experiments to operate the servomechanical system at as low a speed as sustainable and subsequently verify if the assumption holds after δ is obtained (Section 4.2.3).

4.2.3.2 High-Velocity Mode: Coulomb and Viscous Friction Identification

When a servomechanical system operates in the high-velocity mode, the dominant friction components influencing the motion are the Coulomb and viscous friction components. From Figure 4.3, the two-parameter friction model has been used as a good approximation of the four-parameter model. The second phase of the experiment will aim to extract the two parameters of this model. The intersection f_0 of line l_2 and the f-axis is computed as $f_0 = f_2 - \delta f_3$. Thus, the friction model is expressed as

$$f = [f_0 + f_3|\dot{x}|]\text{sgn}(\dot{x}) \tag{4.16}$$

Once f_0 and f_3 are determined, the remaining parameters f_2 and δ are related by $f_2 = f_3\delta + f_0$.

Similar to the procedures depicted in the last section, the DF of the equivalent relay (N_{ER}) is simply the sum of the individual DFs due to the feedback relays (N_{FR1}, N_{FR2}) and the inherent system relay (N_{SR}), i.e., $N_{ER} = N_{FR1} + N_{FR2} + N_{SR}$, where $N_{FR1}(A) = 4h_1/(\pi A)$, $N_{FR2}(A) = -4jh_2/(\pi A)$, and $N_{SR}(A, \omega) = j(4f_0/\pi A + \omega f_3)$. Thus,

$$N_{ER}(A, \omega) = 4h_1/(\pi A) + j[4(f_0 - h_2)/(\pi A) + \omega f_3] \tag{4.17}$$

By varying h_1 or h_2, two relay experiments are conducted, yielding three explicit formulas from which the unknown time constant τ, generalized Coulomb friction f_0, and viscous friction f_3 are computed, since the static gain K has already been estimated during the first phase of the experiment.

$$\begin{bmatrix} f_0 \\ f_3 \end{bmatrix} = \begin{bmatrix} 4/(\pi A_1) & \omega_1 \\ 4/(\pi A_2) & \omega_2 \end{bmatrix}^{-1} \begin{bmatrix} 4h_{21}/(\pi A_1) - \omega_1/K \\ 4h_{22}/(\pi A_2) - \omega_2/K \end{bmatrix} \tag{4.18}$$

where h_{ij} denotes the gain of the ith relay in the jth experiment.

From the same data set, the time constant τ_h of the linear dynamics is also estimated in the high-velocity mode.

$$\tau_h = 2K[h_{11}/(A_1\omega_1^2) + h_{21}/(A_2\omega_2^2)]/\pi \tag{4.19}$$

Then, an average value of the time constant τ is computed from τ_l and τ_h as $\tau = (\tau_l + \tau_h)/2$.

It should be noted that an additional step can be taken to improve the estimation accuracy associated with this describing function approach, which assumes a sinusoidal input. The velocity waveform is not sinusoidal generally, but closer to an repeated isosceles triangle waveform in this mode. As shown in Figure 4.6, assuming the amplitude of the position signal is A, and approximating the velocity signal as a repeating isosceles triangle waveform, the amplitude of velocity signal A_v signal is more accurately expressed as

$$A_v \approx 4\omega A/\pi \tag{4.20}$$

For this reason, a correction factor of $4/\pi$ can be multiplied to the second term in N_{SR}, to improve the estimation accuracy of DF, yielding $\bar{N}_{SR}(A, \omega) = j[4f_0/(\pi A) + 4\omega f_3/\pi]$. Thus,

$$\bar{N}_{ER}(A, \omega) = 4h_1/(\pi A) + j[4(f_0 - h_2)/(\pi A) + 4\omega f_3/\pi] \tag{4.21}$$

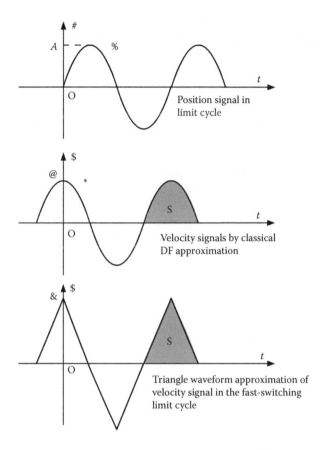

FIGURE 4.6
Improvement of velocity amplitude estimation.

With this correction factor, f_0 and f_3 can be identified as

$$\begin{bmatrix} f_0 \\ f_3 \end{bmatrix} = \begin{bmatrix} 4/(\pi A_1) & 4\omega_1/\pi \\ 4/(\pi A_2) & 4\omega_2/\pi \end{bmatrix}^{-1} \begin{bmatrix} 4h_{21}/(\pi A_1) - \omega_1/K \\ 4h_{22}/(\pi A_2) - \omega_2/K \end{bmatrix} \qquad (4.22)$$

After the second phase, two parameters f_2 and δ are left, but f_2 is computed from $f_2 = f_0 + \delta f_3$, after δ is obtained through an optimization process in Section 4.2.3.

4.2.3.3 Estimating the Boundary Lubrication Velocity by Optimization

The boundary lubrication velocity δ is estimated via an offline optimization process. This will be a single parameter optimization process, since the other friction parameters f_0, f_1, and f_3 are now known and f_2 is a function of δ only too, i.e., $f_2 = f_0 + \delta f_3$.

The harmonic balance condition is rewritten as $N_{ER}(A, \omega) = -1/G_p(j\omega)$, since the reciprocal of $G_p(j\omega)$ is more easily computed than $N_{ER}(A, \omega)$. The objective is to locate a parameter $\hat{\delta}$ that will minimize a performance index

$$J(\hat{\delta}) = \sum_{n=1}^{m} \left\{ \left[\mathbf{Re}(N_{ER}(A_n, \omega_n, \hat{\delta})) + \mathbf{Re}\left(1/G_p(j\omega_n)\right) \right]^2 \right.$$

$$\left. + \left[\mathbf{Im}(N_{ER}(A_n, \omega_n, \hat{\delta})) + \mathbf{Im}\left(1/G_p(j\omega_n)\right) \right]^2 \right\} \qquad (4.23)$$

where m is the total number of data sets from the relay experiments. The optimization process will sweep δ over a range and identify the optimal δ as the value that minimizes J. From Figure 4.3, a bound is further fixed for δ as $0 < \delta < \delta_u$, where $\delta_u = (f_1 - f_0)/f_3$. Compared to the estimation of four parameters concurrently via optimization as in [10], the single-parameter optimization proposed here, which is done offline on existing data sets, is far more efficient and reliable.

4.2.4 Simulation

To elaborate the modeling phases systematically and to highlight the accuracy achievable, consider a servomechanical system described as $G_p(s) = 10/[s(0.2685s + 1)]$ with the four friction parameters given by $f_1 = 0.6$, $f_2 = 0.5$, $f_3 = 0.01$, and $\delta = 0.1$.

4.2.4.1 *Limit Cycle Variation with Relay Gains*

This subsection will highlight how the limit cycle oscillations in the system can vary with different choice of relay gains, and how the guidelines given by the properties of the relay in Section 4.2.3 can be used to position the two phases of the relay experiments in the proper velocity range.

The algorithm is verified through the simulation results as shown in Figure 4.7. Comparing Figure 4.7(a) with Figure 4.7(b), it is observed that when h_1 increases, A decreases and ω increases, while h_2 behaves in the opposite manner. Similarly, Figure 4.8(a, b) shows the validity of relay; i.e., the oscillation amplitude of velocity is invariant with h_1, but it increases with h_2.

Moreover, four sets of relay gains are selected to show four different scenarios as depicted in Figure 4.9(a–d).

In the first scenario as depicted in Figure 4.9(a), $h_2 < f_1$ and no sustainable limit cycle oscillation occurs.

Figure 4.9(b) shows the scenario when the gains of the relay are sufficiently small to maintain $A_v < \delta$. In this case, the output signal $x(t)$ exhibits a relatively fast switching phenomenon, and it has a sinusoidal waveform. The velocity signal has a triangular-shaped periodic waveform.

Figure 4.9(c) presents the scenario when the gains of the relay are still kept small, but now $A_v > \delta$, and the frequency of the waveform $x(t)$ decreases significantly while its amplitude increases significantly. Now, $x(t)$ has a triangular-shaped periodic waveform, while $\dot{x}(t)$ resembles a pulse train.

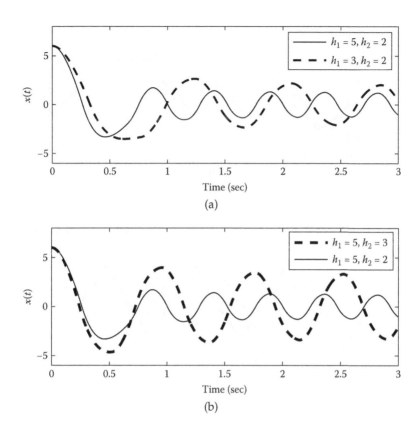

FIGURE 4.7
Investigation of limit cycles of $x(t)$ with choices of different relay gains.

Figure 4.9(d) shows the scenario when the gains of the relay become relatively larger and $A_v > \delta$. The limit cycle becomes fast switching again, and the velocity waveform has recovered the triangular-shape waveform.

The second scenario corresponds to the first phase of the modeling experiment. The fourth scenario corresponds to the second phase of the modeling experiment. From the velocity diagrams, it also shows that the velocity waveforms are more similar to isosceles triangle waveforms in these two scenarios. Thus, (4.22) will give better estimation results than (4.18).

4.2.4.2 Phase 1: Low-Velocity Mode

Following the tuning procedures proposed in Section 4.2.3, by choosing $h_{11} = 0.01$, $h_{21} = 0.605$, $h_{12} = 0.01$, and $h_{22} = 0.603$, the position signals obtained fall in the second scenario, and it yields $\omega_1 = 8.763$, $A_1 = 7.58 \times 10^{-3}$ and $\omega_2 = 14.06$, $A_2 = 2.599 \times 10^{-3}$. By (4.13)~(4.15), the static friction parameter is correctly identified as $\hat{f}_1 = 0.6001$, while the linear dynamics parameters are identified as $\hat{K} = 9.9382$ and $\hat{\tau}_l = 0.2399$.

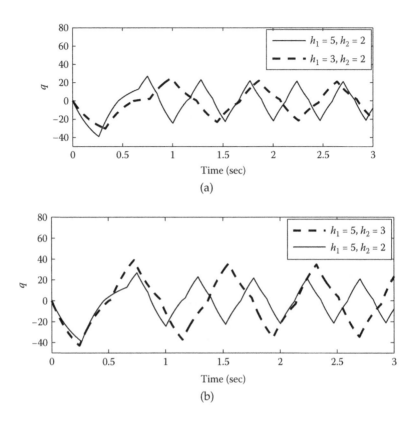

FIGURE 4.8
Investigation of limit cycles of $\dot{x}(t)$ with choices of different relay gains.

4.2.4.3 Phase 2: High-Velocity Mode

In this phase, both DCR gains h_1 and h_2 should be large enough to ensure that the velocity is higher than the boundary lubrication velocity δ, as well as to keep the oscillation frequency sufficiently high. Choosing $h_{11} = 5, h_{21} = 3$, $h_{12} = 3$, and $h_{22} = 2$, two relay experiments are conducted, yielding $\omega_1 = 8.5023$, $A_1 = 3.23$ and $\omega_2 = 8.5486$, $A_2 = 1.935$. Through (4.19) and (4.18), the parameters are successfully identified as $\hat{\tau}_h = 0.2714$, $\hat{f}_0 = 0.4853$, and $\hat{f}_3 = 0.0166$. The estimation of f_3 can be further improved by applying (4.22) rather than (4.18), yielding $\hat{f}_3 = 0.0129$. The final estimation of the time constant is $\hat{\tau} = (\hat{\tau}_l + \hat{\tau}_h)/2 = 0.2557$.

4.2.4.4 Estimation of δ via Optimization

The boundary lubrication velocity δ is identified using the optimization method discussed in Section 4.2.3. Six sets of relay experiment data are used with the system operating in both the low- and high-velocity modes. The bounds for δ are worked out to be within $(0, (\hat{f}_1 - \hat{f}_0)/\hat{f}_3)$, i.e., $(0, 8.89)$.

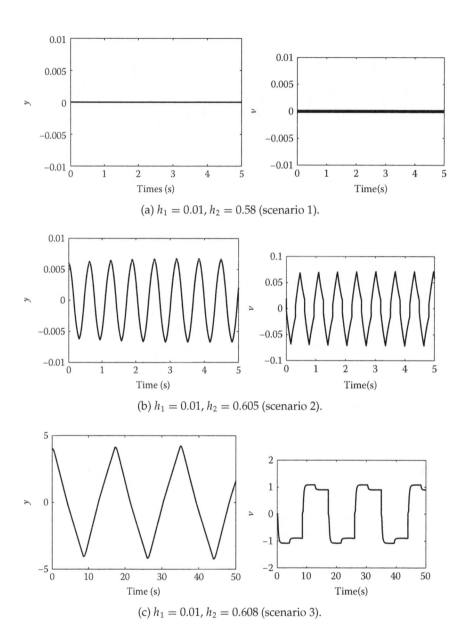

(a) $h_1 = 0.01$, $h_2 = 0.58$ (scenario 1).

(b) $h_1 = 0.01$, $h_2 = 0.605$ (scenario 2).

(c) $h_1 = 0.01$, $h_2 = 0.608$ (scenario 3).

FIGURE 4.9
Four limit cycle scenarios w.r.t. different choices of relay gains. (*continued*)

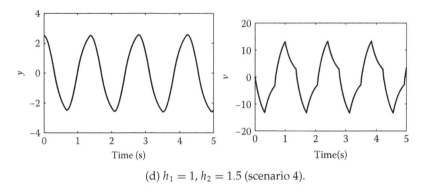

(d) $h_1 = 1$, $h_2 = 1.5$ (scenario 4).

FIGURE 4.9
(*continued*) Four limit cycle scenarios w.r.t. different choices of relay gains.

Using the optimization method discussed in Section 4.2.3, the optimal $\hat{\delta}$ within $(0, 8.89)$ which will minimize the loss function $J(\hat{\delta})$ is found. With all other parameters being identified after the two phases, δ is correctly identified as $\delta_{opt} = 0.09$ with the corresponding minimum performance index $J_{min} = 149.87$. It should be noted that this optimization is done offline on existing sets of data, so there is no need to run extensive and additional experiments for this purpose. Finally, after $\hat{\delta}$ is obtained, $\hat{f}_2$ is directly obtained as $\hat{f}_2 = \hat{f}_0 + \hat{f}_3\delta_{opt} = 0.4865$.

The actual and estimated values of parameters are compared in Table 4.2.

4.2.5 Real-Time Experiments

To illustrate the effectiveness of the proposed method, real-time experiments are carried out on a precision 3D Cartesian robotic system [13], as shown in Figure 4.10. Every axis of the robot is driven by a linear electric motor manufactured by Anorad Co, Shirley, New York. The dSPACE control development and rapid prototyping system, in particular, the DS1103 board, is used. dSPACE integrates the whole development cycle seamlessly into a single environment. MATLAB and Simulink are directly used in the development of the dSPACE real-time control system. This experiment aims to identify the

TABLE 4.2

Summary of Parameter Estimation

Parameter	True	Estimated	Error
f_1	0.6000	0.6001	0.017%
f_2	0.5000	0.4865	2.700%
f_3	0.0100	0.0129	29.000%
δ	0.1000	0.0900	10.000%
K	10.0000	9.9382	0.618%
τ	0.2685	0.2557	4.770%

(a)

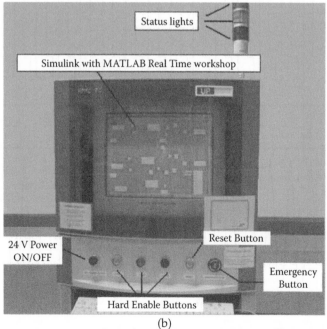

(b)

FIGURE 4.10

Experiment setup. (a) 3D Cartesian robotic system. (b) Computer control platform.

friction parameters of Y-axis servo. For simplicity, the X-axis and Z-axis are fixed on desired positions so that the weight of the loads is evenly distributed on two tracks of the Y-axis, and the disturbance to the Y-axis displacement due to the sliding of the other two axes is negligible.

Several relay experiments are conducted according to the procedures described in Section 4.2.3. The unit of displacement is set to be millimeter. The motor parameters are identified as $K = 579.8480$ and $\tau = 0.6794$, while the friction parameters are identified as $f_1 = 0.3067$, $f_2 = 0.2688$, $f_3 = 1.1087 \times 10^{-4}$, and $\delta = 14.5$. Typically, two patterns of oscillation with choices of different relay gains, under influence of static and Coulomb/viscous frictions accordingly, are shown in Figures 4.11 and 4.12, which correspond to scenarios 2 and 4 as discussed in Section 4.2.4. With the model parameters, a

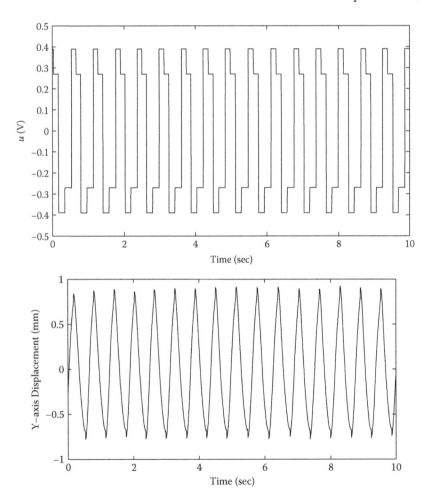

FIGURE 4.11
Input and output signal with $h_1 = 0.06$, $h_2 = 0.33$ (low-velocity mode).

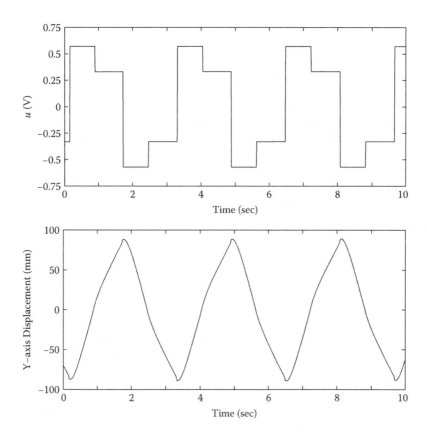

FIGURE 4.12
Input and output signal with $h_1 = 0.12$, $h_2 = 0.45$ (high-velocity mode).

linear feedback controller is commissioned and the feedforward model-based friction compensator is properly initialized as illustrated in Figure 4.13. Since the tracking trajectory is time-varying sinusoidal, and the system itself is type 1, an integral controller is not necessary. By selecting controller parameters as $k_p = 0.005$ and $k_d = 0.001$, Figure 4.14(b) shows the tracking error under the feedback-feedforward control scheme to a sinusoidal reference $r(t) = 50 \sin t$ (unit in mm). For a fair comparison, the tracking performance under the same feedback controller but without a feedforward friction compensator is shown in Figure 4.14(a). And it can be concluded that the tracking performance under normal linear feedback controller is not satisfactory, since under the effects of friction, the linear controller cannot cope with bidirectional, time-varying trajectory well. With adding in the model-based friction compensator, the maximum tracking error is reduced from 0.3 mm to 0.06 mm. Clearly, a significant improvement in reduction of the tracking error is achieved with the friction compensator. The remaining error may be due to ripple forces and other unmodeled uncertainties in the linear motor.

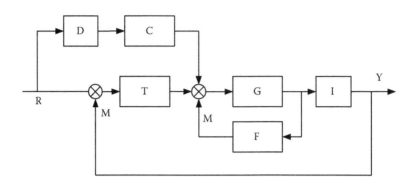

FIGURE 4.13
PID controller with friction precompensator.

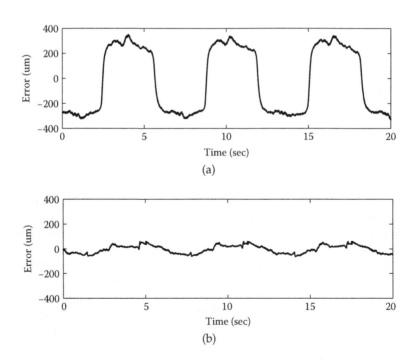

FIGURE 4.14
Closed-loop tracking performance. (a) Without friction compensator. (b) With friction compensator.

4.3 Modeling and Compensation of Ripples and Friction in Permanent Magnet Linear Motors

Permanent magnet linear motors (PMLMs) have the two major nonlinear phenomena: force ripples and friction. Force ripples are strong, position-dependent forces arising from the magnetic structure of a PMLM. The two primary components of the force ripple are the cogging (or detent) force and the reluctance force. The cogging force arises as a result of the mutual attraction between the magnets and iron cores of the translator [6]. These nonlinearities may degrade the system performance. Therefore, a reduction of these effects is of paramount importance if high-speed and high-precision motion control is to be achieved. This section aims to model and compensate force ripples and friction in a typical PMLM, with the assistance of hysteretic relay feedback.

4.3.1 Overall PMLM Model

The nonlinear forces f_{nl} in PMLM are represented as

$$f_{nl} = f_{ripp}(x) + f_{fric}(\dot{x}) + f_{res}(t) \qquad (4.24)$$

where f_{fric} and f_{ripp} represent the friction and force ripple accordingly; f_{res} can be considered to be any other residual forces not considered, possibly arising from model uncertainty and system disturbances present. It is assumed that f_{res} is much smaller than f_{fric} and f_{ripp}, so that it can be ignored.

The frictional force f_{fric} is represented by Coulomb and viscous friction components (with friction model parameters $\bar{f}_1$ and $\bar{f}_2$)

$$f_{fric} = \bar{f}_1 \text{sgn}(\dot{x}) + \bar{f}_2 \dot{x} \qquad (4.25)$$

The force ripple f_{ripp} is represented by a single dominant spatial frequency Ω sinusoidal function with phase shift ϕ.

$$f_{ripp} = C \sin(\Omega x + \phi) = \bar{C}_1 \cos(\Omega x) + \bar{C}_2 \sin(\Omega x) \qquad (4.26)$$

In addition, since the electrical time constant is much smaller than the mechanical one, the dynamics due to electrical induction is omitted. Thus, the following equation describing the final model can be obtained:

$$\ddot{x} = -\left(\frac{K_e K_f + \bar{f}_2 R}{RM} \right) \dot{x}$$

$$+ \frac{K_f}{RM} \left[u - \frac{\bar{f}_1 R}{K_f} \text{sgn}(\dot{x}) \frac{\bar{C}_1 R}{K_f} \cos(\Omega x) - \frac{\bar{C}_2 R}{K_f} \sin(\Omega x) \right] \qquad (4.27)$$

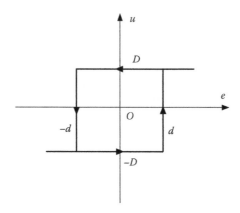

FIGURE 4.15
The hysteretic relay used for identification.

Set $a = (K_e K_f + R\bar{f}_2)/(RM)$, $b = K_f/(RM)$, $f = \bar{f}_1 R/K_f$, $C_1 = \bar{C}_1 R/K_f$, and $C_2 = \bar{C}_2 R/K_f$. Furthermore, introduce

$$\tilde{u} = u - f\,\mathrm{sgn}(\dot{x}) - C_1 \cos(\Omega x) - C_2 \sin(\Omega x) \tag{4.28}$$

so that the linear portion can be written as the following transfer function:

$$G(s) = X(s)/\tilde{U}(s) = b/[s(s+a)] \tag{4.29}$$

In this chapter, an intentional hysteretic relay feedback apparatus is added to induce oscillations from which to identify the system parameters, as shown in Figure 4.15. The hysteretic relay is defined by [15] as

$$u = \begin{cases} D & \text{if } e > d, \text{ or } (e \geq -d \text{ and } u(t_-) = D) \\ -D & \text{if } e < -d, \text{ or } (e \leq d \text{ and } u(t_-) = -D) \end{cases} \tag{4.30}$$

where $e = -x$ under the assumption of a zero reference input, without loss of generation. The full model, in a block diagram form, is illustrated in Figure 4.16.

4.3.2 Model Identification

In this section, the approach to identify the parameters associated with the full model presented in (4.28) and (4.29) will be elaborated. First, an equivalent block diagram model of PMLM will be presented, which segregates cleanly the linear and nonlinear parts of the model. DIDF will be used to approximately describe each of the nonlinear components in the block diagram, and subsequently combined into an overall DIDF. Then, with a harmonic balance analysis, explicit equations to obtain all the model parameters from resultant oscillations will be provided.

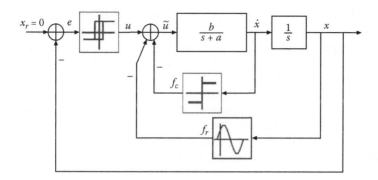

FIGURE 4.16
PMLM under hysteretic relay feedback.

4.3.2.1 Dual-Input Describing Function (DIDF) for Nonlinear Portion of PMLM Model

Since $e = -x$, (4.28) can be written as

$$\tilde{u} = u + f\,\mathrm{sgn}(\dot{e}) - C_1 \cos(\Omega e) + C_2 \sin(\Omega e) \qquad (4.31)$$

In other words, the overall system shown in Figure 4.16 can be converted to the equivalent form of Figure 4.17, so that the linear portion and nonlinear portion are cleanly segregated to facilitate subsequent harmonic balancing for parameter estimation.

In the equivalent system, the system nonlinearities, as well as the intentional relay, all use the error signal as the input, similar to [7]. In general, due to

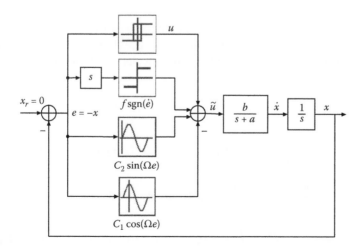

FIGURE 4.17
Equivalent block diagram.

nonzero phase ϕ of force ripple in (4.26), there exists an even nonlinearity in the form of the cosine term. This causes asymmetrical oscillation. The simple sinusoidal-input describing function (SIDF), assuming symmetric sinusoidal input, is not able to describe such even nonlinearity. Thus, the dual-input describing function (DIDF) will be used in this section to approximate each of the nonlinear components. In all of these approximations, a biased sinusoidal input $e(t) = A \sin \omega t + B$ is assumed, where A is amplitude, ω is the oscillating frequency, and B is the bias [9].

Consider first the force ripple nonlinearity $f_{\text{ripp}} = -C_1 \cos \Omega e + C_2 \sin \Omega e$. The following theorem gives the DIDF of this nonlinearity.

Theorem 4.1

The DIDF of nonlinearity $f_{\text{ripp}} = -C_1 \cos \Omega e + C_2 \sin \Omega e$ under biased sinusoidal input $e = A \sin \omega t + B$ is given by $N_{AR}(A, B)$ and $N_{BR}(A, B)$, where

$$N_{AR}(A, B) = \frac{2}{A} [C_1 \sin(\Omega B) + C_2 \cos(\Omega B)] J_1(\Omega A) \qquad (4.32)$$

$$N_{BR}(A, B) = \frac{2}{B} [-C_1 \cos(\Omega B) + C_2 \sin(\Omega B)] J_0(\Omega A) \qquad (4.33)$$

where

$$J_n(z) = \frac{1}{\pi} \int_0^\pi \cos(n\theta - z \sin \theta) d\theta, \qquad n \in \mathbf{N} \qquad (4.34)$$

is a Bessel function of the first kind of order n w.r.t. z.

Proof of Theorem 4.1

First, consider the cosine nonlinearity in the force ripple, i.e., $f_{r1} = C_1 \cos \Omega e$, under biased sinusoidal input $e = A \sin \theta + B$, where $\theta = \omega t$. The general equation for the limit cycle DIDF for a memoryless nonlinearity yields

$$N_{AC}(A, B) = \frac{C_1}{\pi A} \int_0^{2\pi} \cos [\Omega(A \sin \theta + B)] \sin \theta \, d\theta$$

$$= \frac{C_1}{\pi A} \int_0^\pi [\cos(\Omega A \sin \theta + \Omega B) - \cos(\Omega A \sin \theta - \Omega B)] \sin \theta \, d\theta$$

$$= \frac{2C_1 \sin(\Omega B)}{\pi A} \int_0^\pi \sin(\Omega A \sin \theta) \sin \theta \, d\theta$$

$$= \frac{C_1 \sin(\Omega B)}{\pi A} \int_0^\pi [\cos(\Omega A \sin \theta + \theta) - \cos(\Omega A \sin \theta - \theta)] d\theta$$

$$= -2C_1 \sin(\Omega B) J_1(\Omega A)/A \qquad (4.35)$$

The DIDF relating to the bias term is given by

$$
N_{BC}(A, B) = \frac{1}{2\pi B} \int_0^{2\pi} C_1 \cos[\Omega(A\sin\theta + B)]\, d\theta
$$

$$
= \frac{C_1}{2\pi B} \int_0^{\pi} [\cos(\Omega A\sin\theta + \Omega B) + \cos(\Omega A\sin\theta - \Omega B)]\, d\theta
$$

$$
= \frac{C_1 \cos(\Omega B)}{\pi B} \int_0^{\pi} \cos(\Omega A\sin\theta)\, d\theta
$$

$$
= C_1 \cos(\Omega B) J_0(\Omega A)/B \tag{4.36}
$$

Next, consider the sine nonlinearity in the force ripple, i.e., $f_{r2} = C_2 \sin\Omega e$, under biased input $e = A\sin\theta + B$, where $\theta = \omega t$. Similar to the cosine nonlinearity, it yields

$$
N_{AS}(A, B) = \frac{C_2}{\pi A} \int_0^{2\pi} \sin[\Omega(A\sin\theta + B)]\sin\theta\, d\theta
$$

$$
= \frac{C_2}{\pi A} \int_0^{\pi} [\sin(\Omega A\sin\theta + \Omega B) + \sin(\Omega A\sin\theta - \Omega B)]\sin\theta\, d\theta
$$

$$
= \frac{2C_2 \cos(\Omega B)}{\pi A} \int_0^{\pi} \sin(\Omega A\sin\theta)\sin\theta\, d\theta
$$

$$
= \frac{C_2 \cos(\Omega B)}{\pi A} \int_0^{\pi} [\cos(\theta - \Omega A\sin\theta) - \cos(-\theta - \Omega A\sin\theta)]\, d\theta
$$

$$
= 2C_2 \cos(\Omega B) J_1(\Omega A)/A \tag{4.37}
$$

Similarly, for the bias term,

$$
N_{BS}(A, B) = \frac{1}{2\pi B} \int_0^{2\pi} C_2 \sin[\Omega(A\sin\theta + B)]\, d\theta
$$

$$
= \frac{C_2 \sin(\Omega B)}{2\pi B} \int_0^{\pi} [\sin(\Omega A\sin\theta + \Omega B) - \sin(\Omega A\sin\theta - \Omega B)]\, d\theta
$$

$$
= \frac{C_2 \sin(\Omega B)}{\pi B} \int_0^{\pi} \cos(\Omega A\sin\theta)\, d\theta
$$

$$
= C_2 \sin(\Omega B) J_0(\Omega A)/B \tag{4.38}
$$

Noting that $N_{AR} = -N_{AC} + N_{AS}$ and $N_{BR} = -N_{BC} + N_{BS}$, Theorem 4.1 is proofed.

In Theorem 4.1, notice that B is the bias of the error signal, which has an opposite sign to the bias of the position signal x.

For the hysteretic relay given by (4.30) under the biased sinusoidal input $e = A\sin\omega t + B$, the DIDF is given by

$$N_{AH}(A, B) = \frac{2D}{\pi A}\varpi(d, A, B) - j\frac{4Dd}{\pi A^2} \qquad (4.39)$$

$$N_{BH}(A, B) = \frac{D}{\pi B}\upsilon(d, A, B) \qquad (4.40)$$

where

$$\varpi(d, A, B) = \sqrt{1 - \left(\frac{d+B}{A}\right)^2} + \sqrt{1 - \left(\frac{d-B}{A}\right)^2} \qquad (4.41)$$

$$\upsilon(d, A, B) = \sin^{-1}\frac{d+B}{A} - \sin^{-1}\frac{d-B}{A}. \qquad (4.42)$$

For the Coulomb friction nonlinearity $f_c(\dot{e})$ given by $f_c = f\,\mathrm{sgn}(\dot{e})$, with the biased sinusoidal input $e = A\sin\omega t + B$, the DIDF is given by

$$N_{AF}(A, \omega) = 4jf/(\pi A) \qquad (4.43)$$

$$N_{BF} = 0 \qquad (4.44)$$

Here, the bias constant B is eliminated by the differentiator. Thus, the DIDF is equal to the SIDF as discussed in [7].

The overall DIDF of the nonlinear portion is thus given by

$$N_A = N_{AH} + N_{AR} + N_{AF} \qquad (4.45)$$

$$N_B = N_{BH} + N_{BR} \qquad (4.46)$$

4.3.2.2 Parameter Estimation from Harmonic Balance

Note that the linear portion of the PMLM model is a type 1 system. Thus, under the assumption of the existence of a biased sinusoidal limit cycle oscillation, the harmonic balance condition is given by [9] as

$$N_A(A, B, \omega)\,G(j\omega) = -1 \qquad (4.47)$$

$$N_B(A, B, \omega) = 0 \qquad (4.48)$$

With (4.47)~(4.48), together with (4.32)~(4.46), the following three equalities can be established:

$$4Dd/(\pi A^2) = \omega\alpha + 4f/(\pi A) \qquad (4.49)$$

$$-2D\varpi/\pi = -A\omega^2\beta + 2\sin(\Omega B)J_1(\Omega A)C_1 + 2\cos(\Omega B)J_1(\Omega A)C_2 \qquad (4.50)$$

$$-D\upsilon/[\pi J_0(\Omega A)] = -\cos(\Omega B)C_1 + \sin(\Omega B)C_2 \qquad (4.51)$$

where $\alpha = a/b$ and $\beta = 1/b$. Define A_j as the value of A obtained from the jth experiment, so as D_j, d_j, ϖ_j, v_j, B_j, and ω_j. Since five parameters are required to be identified, but only three equations are available, a minimum of two sets of relay experiments is required, which can be obtained by varying the hysteretic relay parameters.

From (4.49), f and α are identified by

$$\alpha = \frac{4(D_1 d_1 A_2 - D_2 d_2 A_1)}{\pi A_1 A_2 (\omega_1 A_1 - \omega_2 A_2)} \tag{4.52}$$

$$f = \frac{A_1^2 \omega_1 D_2 d_2 - A_2^2 \omega_2 D_1 d_1}{A_1 A_2 (\omega_1 A_1 - \omega_2 A_2)} \tag{4.53}$$

C_1 and C_2 can be identified from (4.51) as

$$C_1 = \frac{J_0(\Omega A_2) \sin(\Omega B_2) D_1 v_1 - J_0(\Omega A_1) \sin(\Omega B_1) D_2 v_2}{\pi \sin[\Omega(B_2 - B_1)] J_0(\Omega A_1) J_0(\Omega A_2)} \tag{4.54}$$

$$C_2 = \frac{J_0(\Omega A_2) \cos(\Omega B_2) D_1 v_1 - J_0(\Omega A_1) \cos(\Omega B_1) D_2 v_2}{\pi \sin[\Omega(B_2 - B_1)] J_0(\Omega A_1) J_0(\Omega A_2)} \tag{4.55}$$

β is identified from (4.50) as

$$\beta = \sum_{j=1}^{2} \left[D_j \varpi_j + \pi \sin(\Omega B_j) J_1(\Omega A_j) C_1 + \pi \cos(\Omega B_j) J_1(\Omega A_j) C_2 \right] / (\pi A_j \omega_j^2) \tag{4.56}$$

Thus, a and b are finally obtained by $a = \alpha/\beta$ and $b = 1/\beta$.

4.3.2.3 *Extraction of Frequency Components from DFT*

Compared with the methods proposed in [7,10,14], the new method induces a bias term B in the limit cycle due to the presence of an even nonlinearity with respect to the input signal. This bias arises from asymmetry in the limit cycle. Thus, the amplitude and bias of the oscillation may not be directly obtained from the oscillation accurately, especially when the asymmetric is severe. Instead, discrete Fourier transform (DFT) [11] can be applied to obtain the fundamental frequency, based on which A and B can be extracted.

Without loss of generality, a peak-to-peak N-samples segment $e_s(n)$ consisting of an exact m period of limit cycles is taken from $e(t)$, so that the spectrum leakage is avoided. The bias B is estimated as the mean value of the periodic segment $e_s(n)$ of $e(t)$, or equivalently, the spectral component $E(0)$ divided by the sample size N,

$$B = \frac{1}{N} \sum_{n=0}^{N-1} e_s(n) = \overline{e_s(n)} = E(0)/N \tag{4.57}$$

Since m periods of signal segment are available, the amplitude A of fundamental frequency component is estimated as twice the real part of mth spectral components normalized by the sample size N,

$$
\begin{aligned}
A &= \frac{1}{N} \sum_{n=0}^{N-1} e_s(n) \left[\exp\left(-j\frac{2\pi nm}{N} \right) + \exp\left(j\frac{2\pi nm}{N} \right) \right] \\
&= \frac{2}{N} \sum_{n=0}^{N-1} e_s(n) \cos\left(\frac{2\pi nm}{N} \right) \\
&= 2\overline{e_s(n) \cos\left(2\pi nm/N \right)} \\
&= 2 \, \mathbf{Re}\left[E(m) \right] / N
\end{aligned} \tag{4.58}
$$

4.3.3 Simulation

Consider the PMLM model of (4.28)~(4.29), with parameters set as

$$
a = 4,\ b = 40,\ \Omega = 0.2\pi,\ f = 0.4,\ C = 1 \tag{4.59}
$$

The sampling interval of simulation is fixed as 0.1 ms. This section will first highlight how the limit cycle oscillations are affected with different position phase shift ϕ. By choosing $\phi = 0$ and $\phi = \pi/6$, biased and unbiased limit cycles are observed in Figures 4.18 and 4.19, accordingly, with the same fixed relay parameters $d = 1.2$ and $D = 5$. When $\phi = 0$, there is no cosine term in the model of ripple nonlinearity, and the limit cycle is symmetric with period T. When $\phi = \pi/6$, nonodd ripple nonlinearity $f_r = \sin(0.2\pi x + \pi/6)$ is present, the duty time T^+ and T^- of high and low values of the relay output are not equal, and biased limit cycle oscillations occur.

In the following part, in order to verify the effectiveness of the new method, the parameters of hysteretic relay are chosen as $d_1 = 1.2$, $D_1 = 5$, $d_2 = 0.8$, and $D_2 = 3$, to identify the system model of (4.28)~(4.29), with the same parameter set as in (4.59), with $\phi = \pi/6$. The simulation results for the two limit cycles within five periods are shown in Figures 4.19 and 4.20 accordingly. The asymmetry and bias in the oscillation are evident in these figures.

DFT is first applied to complete cycles of $e(t)$, beginning from when $e(t)$ is at maximum value and ending five complete periods later. The spectrum of the signals is shown in Figure 4.21. A and B can be obtained from (4.57)~(4.58). In this way, the frequency and amplitude of the fundamental harmonics, and DC bias are obtained as $\omega_1 = 10.2834$, $A_1 = 2.4639$, $B_1 = 0.1222$, $\omega_2 = 10.2099$, $A_2 = 1.4819$, and $B_2 = 0.1763$. With the explicit equations given in (4.52)~(4.56), the system parameters are correctly identified as $a = 4.0089$, $b = 39.4076$, $C_1 = 0.4423$, $C_2 = 0.8810$, and $f = 0.4107$. Table 4.3 also shows that the error of estimation can be kept to about 10% and below, which demonstrates the efficiency and applicability of the proposed method. From (4.54) and (4.55), the formulae of computing C_1 and C_2 only differ from each other by terms of $\sin(\Omega B)$ and $\cos(\Omega B)$. The error of estimation of C_1 is larger

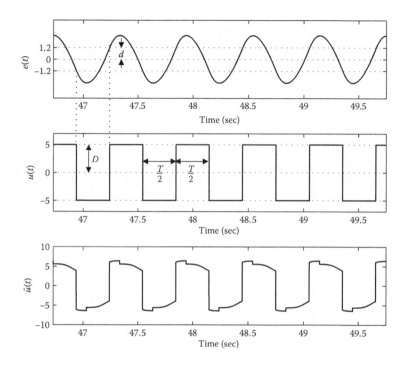

FIGURE 4.18
Input $e(t)$ and output $u(t)$ of the hysteretic relay and actual control signal $\tilde{u}(t)$ with $d = 1.2$, $D = 5$, and $\phi = 0$.

than C_2 since the gradient of $\sin(\Omega B)$ is steeper than that of $\cos(\Omega B)$, when ΩB is relatively small.

4.3.4 Real-Time Experiments

To illustrate the effectiveness of the proposed method, real-time experiments are conducted on a PMLM at the Singapore Institute of Manufacturing Technology (SIMTech), as shown in Figure 4.22, using the dSPACE Alpha Combo multiprocessor control system with MATLAB Simulink Real-Time Workshop. In this dual-DSP system, the dSPACE DS1004 DSP board is used for computationally intensive tasks associated with execution of control algorithms, while the DS1003 DSP board is able to deal efficiently with all the necessary input-output (I/O) tasks. Both boards are real-time interface enabled and configured to give optimal performance via the decentralization. The overall block diagram of the Simulink program used in this experiment is shown in Figure 4.23.

4.3.4.1 Identification of the Spatial Cogging Frequency

In the first part of the experiment, the spatial cogging frequency Ω is identified from the velocity curve with a step voltage input to the PMLM. Figures 4.24

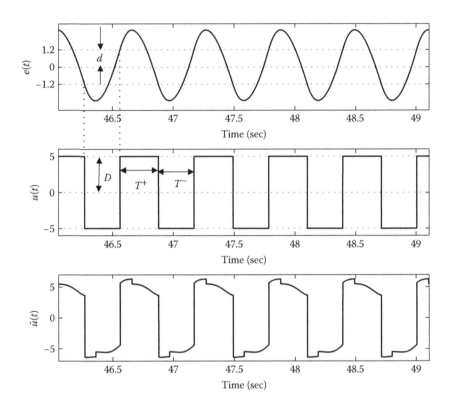

FIGURE 4.19
Input $e(t)$ and output $u(t)$ of the hysteretic relay and actual control signal $\tilde{u}(t)$ with $d = 1.2$, $D = 5$, and $\phi = \pi/6$.

and 4.25 show the open-loop response of the PMLM with different input voltages. From these two figures, after the initial transience, an almost constant velocity trend is observed in the position signals. However, due to the existence of force ripples, the actual velocity signal manifests a periodic oscillating behavior about a mean level. Denote the mean velocity in the steady state as $\bar{v}$ and the period of the velocity oscillation as T_v, then the spatial cogging frequency (in rad/m) can be simply measured as

$$\Omega = \frac{2\pi}{\bar{v}T_v} \tag{4.60}$$

From Figure 4.24, $T_v = 0.2744\,$s and $\bar{v} = 0.18388\,$m/s. Hence, by (4.60), Ω is estimated as 124.39 rad/m. Similarly, from Figure 4.25, $T_v = 0.1620\,$s, $\bar{v} = 0.3154\,$m/s, and Ω is obtained as 122.97 rad/m. Thus, the spatial cogging frequency is identified as $\Omega = 123.68\,$rad/m by taking the mean value of the above two.

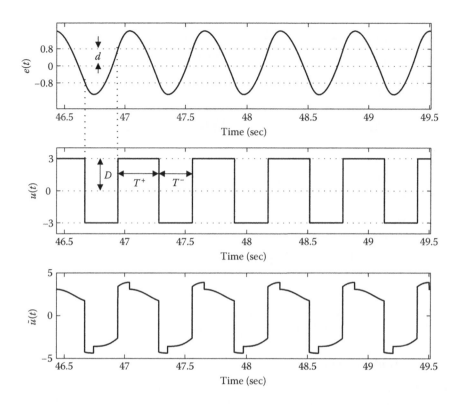

FIGURE 4.20
Input $e(t)$ and output $u(t)$ of the hysteretic relay and actual control signal $\tilde{u}(t)$ with $d = 0.8$, $D = 3$, and $\phi = \pi/6$.

4.3.4.2 Parameter Estimation

To estimate the model parameters of the selected model of PMLM, choose $d_1 = 5 \times 10^{-4}$ m, $D_1 = 0.6$ V, $d_2 = 8 \times 10^{-4}$ m, and $D_2 = 0.7$ V. The sampling period for experiment is set to 4 ms, and the reference position is set 7 cm from the homing position. The results of inputs and outputs of hysteristic relay over five periods of oscillations are shown in Figures 4.26 and 4.27 accordingly. From these figures, the oscillation frequencies are $\omega_1 = 28.3537$ rad/s and $\omega_2 = 28.560$ rad/s. Figure 4.28 shows the spectrum of the window with five periods of $e(t)$ near the DC region, from which it can be concluded that it is appropriate to approximate the steady oscillating signals by their dominant fundamental frequency components plus DC biases. By (4.57)~(4.58), the limit cycle parameters are obtained as $A_1 = 2.4317 \times 10^{-3}$ m, $B_1 = 1.4552 \times 10^{-3}$ m, $A_2 = 3.4978 \times 10^{-3}$ m, and $B_2 = 1.6905 \times 10^{-3}$ m. With the explicit equations given in (4.52)~(4.56), the system parameters are identified as $a = 6.474$, $b = 4.284$, $C_1 = 0.199$, $C_2 = -0.303$, and $f = 0.042$.

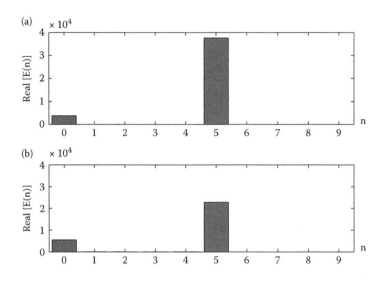

FIGURE 4.21
Spectrums of limit cycles near the DC region with $m = 5$. Left: With $d = 1.2$, $D = 5$, $\phi = \pi/6$, and $N = 29,295$. Right: With $d = 0.8$, $D = 3$, $\phi = \pi/6$, and $N = 32,615$.

4.3.4.3 Model Compensation

To verify the model obtained, a linear feedback controller is commissioned and the feedforward model-based nonlinear compensator is initialized as shown in Figure 4.29. The desired moving profile is set as $x_d = 0.02\sin(2\pi t)$ (unit in m). After fine-tuning, the linear proportional-integral-derivative (PID) feedback controller is set as $k_p = 0.0243\,\text{V}/\mu\,\text{m}$, $k_i = 0$, and $k_d = 0.00013\,\text{V}$ s/μm, with the feedforward controller settings based on the parameters estimated earlier; the tracking errors are shown in Figure 4.30(a). For a fair comparison, the tracking performance is tested with the same linear feedback controller, but without the ripple and friction compensation. The results are shown in Figure 4.30(b). By comparing the above two compensation results, We are able to observe that the maximum tracking errors are reduced dramatically from 15μ m to less than 4.7μm (or around 70% improvement), which also verifies the validity of the model parameters obtained in Section 4.3.4.

TABLE 4.3

Summary of Simulation Results

Parameter	Actual	Estimated	Error %
a	4.0	4.0089	0.22
b	40.0	39.4076	−1.48
C_1	0.5000	0.4423	−11.54
C_2	0.8660	0.8810	1.73
f	0.4	0.4107	−2.67

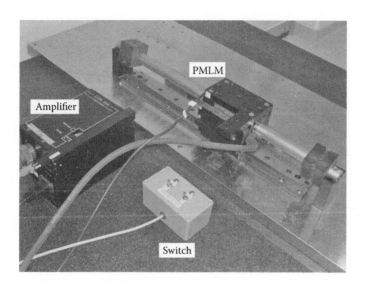

FIGURE 4.22
The PMLM used in this experiment.

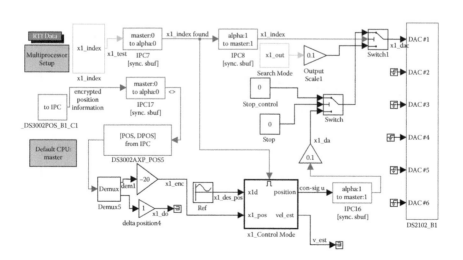

FIGURE 4.23
The Simulink program for experiment.

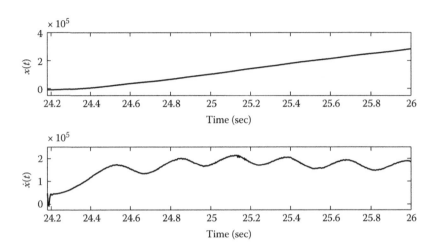

FIGURE 4.24
Position $x(t)$ (in μm) and velocity $\dot{x}(t)$ (in μm/s) of the PLMM with $u = 0.3$ V.

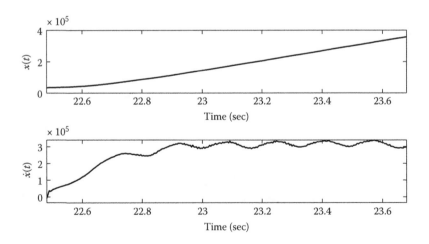

FIGURE 4.25
Position $x(t)$ (in μm) and velocity $\dot{x}(t)$ (in μm/s) of the PLMM with $u = 0.5$ V.

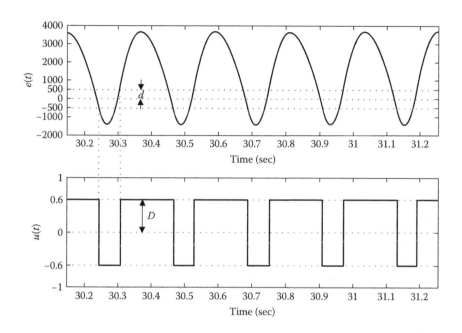

FIGURE 4.26
Input $e(t)$ (in μm) and output $u(t)$ (in V) of the hysteretic relay with $d = 0.5$ mm, $D = 0.6$ V.

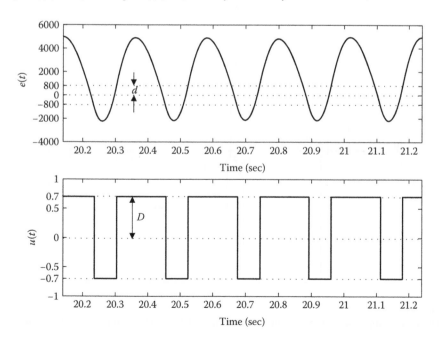

FIGURE 4.27
Input $e(t)$ (in μm) and output $u(t)$ (in V) of the hysteretic relay with $d = 0.8$ mm, $D = 0.7$ V.

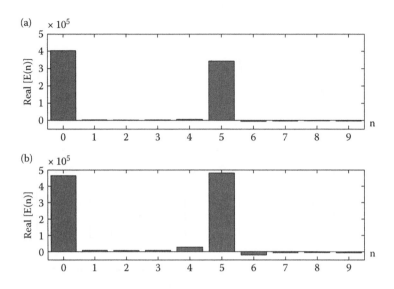

FIGURE 4.28
Spectrums of limit cycles near the DC region with $m = 5$. Left: With $d = 0.5\,\mathrm{mm}$, $D = 0.6\,\mathrm{V}$, and $N = 277$. Right: With $d = 0.8\,\mathrm{mm}$, $D = 0.7\,\mathrm{V}$, and $N = 275$.

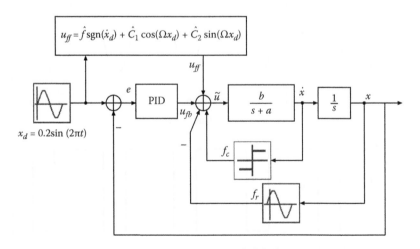

FIGURE 4.29
Design of compensation scheme, with combination of feedback control u_{fb} and feedforward control u_{ff}.

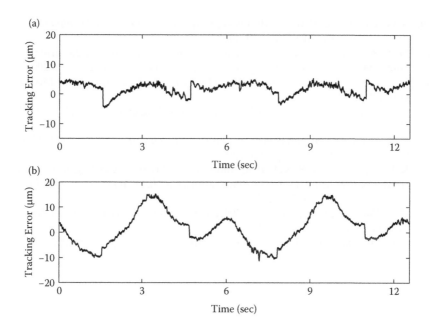

FIGURE 4.30
Tracking performance of control schemes. (a) With nonlinear feedforward compensation. (b) Without nonlinear feedforward compensation.

References

1. K. J. Astrom and B. Wittenmark. *Adaptive control*. 2nd ed. Reading, MA: Addison-Wesley, 1994.
2. K. J. Astrom and T. Hagglund. *Automatic tuning of PID controllers*. Research Triangle Park, NC: Instrument Society of America, 1988.
3. K. J. Astrom and T. Hagglund. Automatic tuning of simple regulators with specifications on phase and amplitude margins. *Automatica*, 20(5), 645–651, 1984.
4. B. Armstrong-Helouvry, P. Dupont, and C. C. de Wit. A survey of models, analysis tools and compensation methods for control of machines with friction. *Automatica*, 30(7), 1083–1138, 1994.
5. K. J. Astrom. Oscillations in systems with relay feedback. In K. J. Astrom, G. C. Goodwin, and P. R. Kumar, editors, *Adaptive control, filtering and signal processing*. Boston: Spring-Verlag, 1995.
6. I. Boldea and S. A. Nasar. *Linear motion electromagnetic devices*, chap. 2–4. NY: Taylor & Francis, 2001.
7. S. L. Chen, K. K. Tan, and S. Huang. Friction modeling and compensation of servomechanical system using a dual-relay feedback approach. *IEEE Transactions on Control Systems Technology*, 17(6), 1295–1305, 2009.
8. M. Friman and K. V. Waller. A two-channel relay for autotuning. *Industrial and Engineering Chemistry Research*, 36, 2662–2671, 1997.

9. A. Gelb and W. E. V. Velde. *Mutiple-input describing functions and nonlinear system design.* New York: McGraw-Hill, 1968.

10. M. S. Kim and S. C. Chung. Friction identification of ball-screw driven servomechanisms through limit cycle analysis. *Mechatronics*, 16, 131–140, 2006.

11. D. K. Linder. *Introduction to signals and systems*, chap. 20. Singapore: McGraw-Hill, 1999.

12. J. J. E. Slotine and W. Li. *Applied nonlinear control.* Englewood Cliffs, NJ: Prentice-Hall, 1991.

13. K. K. Tan, T. H. Lee, and S. Huang. *Precision motion control: Design and implementation.* 2nd ed. London: Springer, 2008.

14. K. K. Tan, T. H. Lee, S. N. Huang, and X. Jiang. Friction modeling and adaptive compensation using a relay feedback approach. *IEEE Transactions on Industrial Electronics*, 48(1), 169–176, 2001.

15. Q.-G. Wang, T. H. Lee, and C. Lin. *Relay feedback: Analysis, identification and control.* London: Springer-Verlag, 2002.

16. C. C. Yu. *Autotuning of PID controllers.* 2nd ed. London: Springer, 2007.

5

Model Predictive Control of Precise Actuators

In the preceding two chapters, identification techniques were applied to obtain the nonlinear dynamics and the controller designed to compensate the effects of these nonlinearities, such as hysteresis, friction, and ripple forces. However, constraints on the precise actuator inputs, states, and outputs are not addressed in these methods. It is well known that the control of a typical actuator often lies at the intersection of constraints. Model predictive control (MPC) has become a standard approach to solve control problems with constraints in a wide range of application. This chapter will present the use of MPC formulation to control a class of precise actuators, called the ultrasonic motor (USM).

The USM is a type of piezoelectric actuator that uses some form of piezoelectric material and relies on the piezoelectric effect. The USM offers advantages of high resolution and speed to ensure precision and repeatability, so it is widely used in precision engineering, robotics, and medical and surgical instruments where high accuracy is required. While a typical piezoelectric actuator (PA) is driven directly by the deformation of the piezoelectric material when a voltage applies, the USM provides motions by virtue of the friction between the piezoelectric material on the stator and the rotor. Thus, the USM offers another advantage of theoretically unlimited travel distance in comparison with the typical piezoelectric actuators. Figure 5.1 shows the internal structure and working principle of the USM. The rotor motion is based on an alumina tip attached to the piezo-ceramic plate (the stator), segmented on one side by two electrodes. Depending on the desired direction of motion, the left or right electrode of the piezo-ceramic plate is excited with a standing wave to produce high-frequency vibration. Because of the asymmetric characteristic of the standing wave, the tip moves along an inclined linear path with respect to the friction bar surface and drives the rotor forward or backward. Each oscillatory cycle of the tip can transfer a 0.3 μm linear movement to the friction bar. With the high-frequency oscillation, it will result in a smooth and continuous rotor motion. An external drive is used to convert analog input signals to the required high-frequency drive signals. The motor is employed in a semiautomated device for medical operations on a human ear tympanic

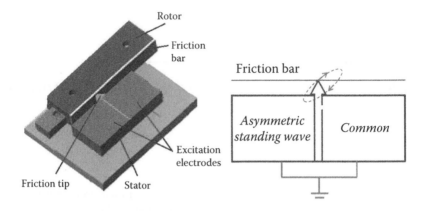

FIGURE 5.1
Linear ultrasonic motor structure and motion description.

membrane. The operating conditions defined by the manufacturer are travel range ±9.5 mm, maximum speed 400 mm/s, and input voltage ±10 V. Because of the strict constraints in medical operation, additional control constraints arise: tracking overshoot less than 5%, settling time within 0.1 s, and maximum steady-state error 0.02 mm.

In the friction-based motion of USM, frictional forces are the main disturbance that degrades the closed-loop performance. Because the friction presents a nonlinear switch that is dependent on the motion direction, using a single linear model to design a linear controller results in inaccuracy, especially at low-speed control [2]. Additionally, a practical controller should respect the physical limitations of the motor input and safety constraints on the system variables (e.g., position range, speed).

The control algorithms reported in the open literature are based on decoupling the friction model from the linear motion system and mitigating it through a nonlinear model-based input beside a linear regulator such as proportional-integral-derivative (PID). They differ from one another in nonlinear friction models as well as the methods to compensate for it. Along this perspective, many research efforts have been focusing on building accurate friction models [7,12,30]. The compensation, usually of bang-bang type in practice, resolves the friction problem and leaves PID with other unmeasured disturbances, including the friction model mismatch. The approaches are simple to implement, and if properly tuned, they provide fast transient response, good static accuracy, and robustness to the motor parameter variations [31]. However, the nonlinear compensation is contingent on asymptotic stability, which relies on the specified friction model. The frictional effects can also depend on rotor position and system degeneration, so a fixed friction model may require more time to compute. Finally, such control tactics do not deal systematically with constraints on the control input and variables, so manual safety considerations have to be addressed.

The MPC control approach is a useful tool for solving constraint problems. Industry experts developed the first description of MPC, which is reported in [35], pointing out significant benefits of a multilevel control. With better dynamic control quality, model-based optimization drives the optimal set point nearer to the constraints to minimize costs. This structure is now followed by many practitioners and researchers [8,20,29]. The flexibility of the MPC framework is that it can use mathematical programming to solve systematically constrained optimization. Model identification and state estimation techniques have been integrated, which are used in most of the current MPC research literature [22,42]. A recent rising approach to deal with friction in electrical drives is based on piecewise affine (PWA) modeling of the nonlinear frictional effects. In [13,41], the authors applied this method on model predictive control to design time-optimal control strategies. Although the method still depends on the choice of friction models, and it does not consider robustness issues, the tracking performance is promising. This chapter will use the PWA model to design a robust MPC via the familiar quadratic programming for a tractable solution. The commonly used friction models are approximated by several linear segments, so MPC is aware of this impeding force at low speed. A constrained optimal control problem for PWA systems is then formulated to provide stability-guaranteed input. Specifically, an integral MPC design imposes the robustness on model-plant mismatch near zero speed. Implementation of the real-time control is handled by a gain scheduling table so that the complexity is comparable to the traditional feedforward PID.

5.1 Model Predictive Control Concepts for Motion Tracking

Originating from linear quadratic regulator (LQR) theory, MPC has picked up the internal model control structure, as shown in Figure 5.2. This leads to two properties: (1) the performance of *unconstrained* MPC is not inherently better than that of classic control, and (2) from an optimal control point of view, optimizing the MPC controller for a certain performance criteria is affine, much simpler than a nonlinear function of the traditional feedback controllers (such as PID) [10]. The former property is easy to understand, since without constraints MPC only results in a linear feedback gain that can be tuned using other methods. However, the latter explains that optimal control, especially with constraints involved, is more advantageous with MPC. Therefore, when mathematical tools such as linear and quadratic programming come into the field, MPC advances along this direction. This new approach was adopted widely for multivariable systems in the industry. In the following subsections we will look closely at how to customize MPC for precise actuators.

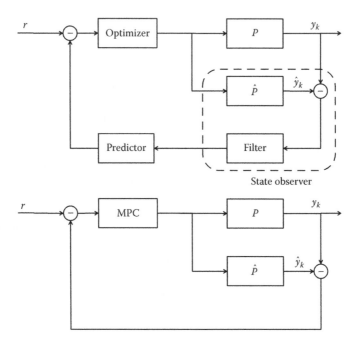

FIGURE 5.2
MPC structure under internal model view.

5.1.1 Prediction and Optimization

MPC has its unique features to stand out in the broad optimization world. Applied on large dynamical systems with constraints, MPC optimizes over a finite-time horizon, thus less expensive and more suitable for control purposes. Using the current dynamic state of a system, MPC calculates a sequence of future optimal inputs to handle it. Only the first optimal input is sent to the system. The idea is that disturbance sources (model uncertainties, disrupting noises) would make other future inputs useless. In the next sampling time, the optimization window moves one step forward, and the procedure repeats. In other words, MPC runs on a feedback manner instead of one-time optimal design. This feature is named *receding horizon control*. Also, because the finite-time optimal control alone does not guarantee the overall system stability as well as LQR does, MPC has added other features for convergence analysis: terminal cost and terminal region [23].

This section describes the adaptation of linear MPC to precise motion systems. Consider the following linear mathematical model of the USM:

$$x(k+1) = Ax(k) + Bu(k)$$
$$y(k) = Cx(k) \qquad\qquad (5.1)$$

subject to the constraints

$$x_{min} \leq x(k) \leq x_{max}$$

$$u_{min} \leq u(k) \leq u_{max} \tag{5.2}$$

where $x(k) \in \mathbb{R}^n$ represents the rotor position and speed, $y(k) \in \mathbb{R}^p$ is the position, and $u(k) \in \mathbb{R}^m$ represents the input voltage delivered to the motor drive.

The MPC simple form can then be posed as an optimization problem:

$$V_N^o(x_0, U) = \min_{U} x_N^T P x_N + \sum_{k=0}^{N-1} \left(x_k^T Q x_k + u_k^T R u_k \right)$$

$$subj. to \quad x_{min} \leq x_k \leq x_{max}, \quad k = 0, ..., N$$

$$u_{min} \leq u_k \leq u_{max}, \quad k = 0, ..., N \tag{5.3}$$

where the state evolution x_k is predicted from (5.1). The decision variable is $U = [u_0^T, ..., u_{N-1}^T]^T$ and the tuning parameters Q, R are constant and positive definite. P is given by the solution of the Riccati equation:

$$P = Q + A^T (P - P B (R + B^T P B)^{-1} B^T P) A \tag{5.4}$$

Equation (5.3) can be recast as a quadratic program in U using a stacked vector and solved repetitively at each time k for the current state $x_0 = x(k)$. Only the first optimal input voltage $u(k) = u_0$ is implemented, and the problem is solved again at the next time interval when a new position measurement becomes available. This approach offers a key feature to take into account the bounds on the position range and rotor speed. In summary, with linear systems MPC results have been well established.

The real motion system of actuators is, however, not linear as the model in (5.1). $u(k)$ has to overcome a frictional force F. Friction affects the model differently at low speed and normal speed, so it introduces different linear models based upon the current state (velocity and input). This accounts for an inconsistent prediction model for MPC if a linear control law is to be designed.

To address this nonlinear modeling, piecewise affine modeling is applied to this class of systems to leverage on the advantage of parametric MPC. The nonlinear uncertainty in low-speed friction will then be taken care of by designing a robust control input for the low-speed subspace, as seen in Section 5.1.2.

5.1.2 Offset-Free and Robust MPC

Among the practical extensions of MPC, the two properties of *offset-free tracking* and *robust control* are the most critical features that concern MPC implementation, whether it is in the original process industry, economics (slow response processes), or new areas such as automotive (fast). They are essentially relevant to the types of disturbances/uncertainties contained inside any controlled systems, for example, modeling mismatch, input and output disturbances. In the next subsections, a review of techniques developed to deal with these problems is discussed.

5.1.2.1 Offset-Free Tracking

Offset-free tracking in process MPC refers to an essential group of techniques that estimate the disturbance infiltrating the model through measured outputs and compensating the offset because of the disturbance by adjusting the MPC target generation. Much research effort has embraced this disturbance estimation and compensation method. The disturbance compensation approach is commonly adopted in the industry with the advantages it offers. Disturbance sources are observed from the process data and the disturbance model is determined. With the model, the disturbance rejection workload is mainly entrusted to a state estimator that leverages on Kalman filters to derive a smooth and optimal estimation of the disturbances. The states and input targets are updated based on the estimated bias so that MPC regulation subject to constraints can run without increased complexity. Over the last two decades, efforts were devoted to refine and expand the applicability of the disturbance model for MPC. In process control, the hallmark paper [26] provided an analytical base for MPC with different disturbance models: zero-mean, output, and input step disturbances. The authors in [25,28] independently introduced a dynamical disturbance model that can be rendered into a statespace representation. They discussed crucial conditions under which offset-free control can be guaranteed. These conditions demand that the augmented system model is detectable and the number of unmeasured disturbances is *equal* to the number of measured outputs (see a proof in [34], Chapter 1). The second requirement was subsequently relaxed to be *less than or equal* [19]. The disturbance estimation approach has been extended to deal with servo disturbance [20], and it is recently explored on nonlinear systems [24]. On the application sides, this method was used in a project to enhance MPC-based software, developed by AspenTech [8].

However, there are clear and outstanding *constraints* associated with this conventional framework. First, from an application and industrial perspective, it achieves the improved margin of performance from disturbance estimation by passing on the complexities of configuring and tuning an accurate dynamic disturbance model to the control engineers. Inaccuracy in the modeling effort will directly incur inadequate compensation and lead to offsets. These complex yet crucial responsibilities are not readily assimilated by practitioners, so the framework cannot be effectively rooted into place in the industry. Second, a general, systematic, and practically amenable way to yield accurate disturbance estimation without undue burden on the practitioners has not appeared so far, largely because of the nongeneral restrictions to be overcome. These restrictions include conditions of the augmented model (plant + disturbance) on detectability and conditions on the rank of the state-to-output matrix [8], which have remained a tall order to generalize straightforwardly. For theorists, the coexistence of the disturbance compensation scheme and the MPC controller in the augmented system has long impeded closed-loop analysis to be done in the same rigorous and comprehensive way when they are separated. The specific choice of a disturbance model has direct implications on the performance of the MPC. Besides, a class of systems will always exhibit performance limitations with this approach, as proven in [3]. Third, established robust and nonlinear system control techniques that are mature in their own rights have been unable to flourish under this tracking framework of MPC mainly because of the frequent target adjustment. To date, there have not been many results on offset-free robust/nonlinear predictive control. Attempts to impose robust control with an output disturbance model [18] or to extend the control to a nonlinear system with a dynamic disturbance model [24] have faced challenging implementation issues. The disturbance estimation framework has long remained as the springboard for tracking MPC designs, but equally long and persistent have been the constraints inherited under the framework. Another drawback, particularly relevant to fast sampling systems such as precise actuators, is the limited computing ability of controllers.

As an alternative, one would prefer using the integrating state variables to track important outputs. Departing from the disturbance compensation framework, it preserves the offset-free property using simply the integration of the output error, tantamount to the integral control action of the PID controller. It is arguably the easiest to understand method in offset-free control and one which can be more readily accepted by practitioners. Integrated states are commonplace in robust tracking control of linear systems [43, Section 14.8 of 44] and nonlinear systems [9,14,38], and they are not sensitive to modeling errors. However, this simple concept has not found its way to tracking MPC, though integrated states have been mentioned, not in depth, in various MPC contexts [1,11,36,39]. Here we propose an augmentation of PID states to facilitate a unified framework to allow tracking MPC and robust feedback control to coexist in a complementary manner.

To get a linear feedback involving proportional-integral-differential gains, it is necessary to form a system state that contains the corresponding variables. Provided it is the case, an optimal linear feedback gain will be also an optimal PID gain. Controllability and observability of such systems are analyzed as well.

Augment (5.5) with an integral of the tracked output $\theta(k) = \sum y(k)$ to ensure zero offset during the steady state. The following PI state model is used:

$$\begin{bmatrix} x(k+1) \\ \theta(k+1) \end{bmatrix} = \begin{bmatrix} A & 0 \\ C & I_q \end{bmatrix} \begin{bmatrix} x(k) \\ \theta(k) \end{bmatrix} + \begin{bmatrix} B \\ 0 \end{bmatrix} u(k)$$
$$y(k) = Cx(k) \tag{5.5}$$

The objective is to design a finite-horizon optimal control based on the augmented system (5.5) so that $y(k)$ tracks a piecewise constant reference.

Proposition 5.1
The PI-augmented system (5.5) is detectable. Further, it is controllable if and only if (A, B) is controllable and

$$rank \begin{bmatrix} A - I_n & B \\ C & 0 \end{bmatrix} = n + q \tag{5.6}$$

Proof The Hautus condition for observability is

$$rank \begin{bmatrix} A^T - \lambda I_n & C^T & C^T \\ 0 & I_q - \lambda I_q & 0 \end{bmatrix} = n + q \quad \text{for all } \lambda \in \mathbb{C} \tag{5.7}$$

The condition (5.7) does not hold only at $\lambda = (1, 0)$, but the unobservable integrating state can be controlled to decay to a constant so the system is detectable.

Similarly, (5.6) follows directly from Hautus controllability where only the case of $\lambda = (1, 0)$ is to check.

From the system detectability, an observer can make use of the system (5.5) to estimate the current state, and simply calculate the integral and differential state through a sum of the estimated $\hat{y}(k) = C\hat{x}(k)$ and its difference.

Since (A, C) is observable, the observer is designed as

$$\hat{x}(k) = A\hat{x}(k-1) + Bu(k-1) + L_x[-y(k-1) + C\hat{x}(k-1)]$$
$$\hat{\theta}(k) = \hat{\theta}(k-1) + C\hat{x}(k-1) + CL_x[-y(k-1) + C\hat{x}(k-1)] \tag{5.8}$$

It is only necessary to design the observer gain L_x as $eig(A + L_xC) < 1$ so that $\hat{x}(k) - x(k) \to 0$. This automatically leads to $\Delta\hat{y}(k)$ being stable. The integral estimation error is not required to decay to zero, but a steady state because $\hat{\theta}(k) - \theta(k) \to const$ means $\hat{y}(k) - y(k) \to 0$.

Apply (5.5) to the MPC formulation in (5.3). Because this augmentation does not increase the size of decision variables or constraints, the optimization complexity stays intact. As seen later in Section 5.2.2, regulating the output error integrating state will ensure offset-free tracking in the position $y(k)$.

5.1.2.2 Robust Formulation

A natural question for model predictive control is its robustness to model uncertainty and noise. The original open-loop model predictive control is about determining the current control action by solving an optimal problem with only the information of the current state. This method cannot restrain the spread of predicted trajectories resulting from disturbances, so feedback policies including disturbance model along the controlled horizon are researched [21,37]. It implies that for a specified range of model variations/noises, the system stability is maintained and the performance specifications are met. The improvement is based on a trade-off between the accuracy of the uncertainty description and the computational complexity of the controller synthesis.

In precise actuator applications, however, the uncertainty mainly lies in friction modeling near zero speed. So, the focus can be solely placed on robust control at low speed instead of along the prediction trajectory. One common method is describing this uncertainty in terms of bounded compact sets and reducing its effect through robust control within a local region. Asymptotic stability of the origin cannot be established, but of this region only. It suggests that a Lyapunov function must be zero within this set and nonzero otherwise [15,37]. However, this is not a necessity if the piecewise affine model is used for MPC, as it is demonstrated in Section 5.2.

The controller designed in this chapter is a dual-mode controller. In actuator control, the uncertainty in friction modeling accounts for the main part of position error. Most often, linear models represent the motion system well at medium and high speed, but fail to describe the low-speed behavior because of the nonlinear friction force. Hence, a piecewise affine model is used here to model the motion system at the two regions in both motion directions. An adaptive MPC gain scheme is then adopted: a normal MPC input is to drive the system state near the desirable position set point, after which a local MPC robust gain specially designed for the nonlinear friction is applied.

Under this framework, an opening is created for a straightforward use of robust control into the core of offset-free MPC. Disturbances that render active constraints at steady state and drive offset-free MPC tracking infeasible [28,34] are now dealt by established robust control (H_∞). Inaccuracy in the disturbance model used in the robust control is in turn mitigated by the offset-free property of the tracking MPC. The disturbance model is no longer necessary and the system state can be predicted through an augmented model (i.e., [32]). This PI-augmented model is the simple yet crucial key under the proposed framework to bridge offset-free MPC to robust control.

5.2 Hybrid MPC for Ultrasonic Motors

5.2.1 Piecewise Affine Model of Motion

Consider a classical linear motion model that takes the form

$$
\begin{bmatrix} \dot{y} \\ \dot{v} \end{bmatrix} = \begin{bmatrix} 0 & 1 \\ a & b \end{bmatrix} \begin{bmatrix} y \\ v \end{bmatrix} + \begin{bmatrix} 0 \\ c \end{bmatrix} (u - F(v)) \tag{5.9}
$$

where y, v is the rotor position and velocity; $F(v)$, the friction as shown in Figure 5.3, consists of the constant Coulomb friction f_c, viscous friction $f_v = kv$, and Stribek effect f_s showing how the friction continuously decreases as the motor accelerates.

From there, the motion system of ultrasonic motor can be represented in four regions, A–D, as in Figure 5.3. Because the viscosity is linear in v, models in regions A and D can be represented by (5.9). In regions B and C, the same structure can be employed to approximate the system, but with different linear dynamics; the complexity in the presliding regime will be addressed by the robust design in Section 5.2.2.

First, the effective relationship from input to rotor position is identified. This addresses the relation between the output position and the input

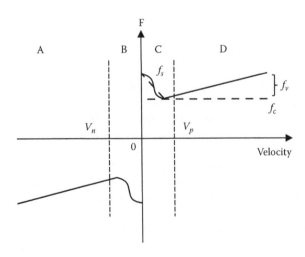

FIGURE 5.3
Motion friction can be described by linear segments over four regions, from A to D.

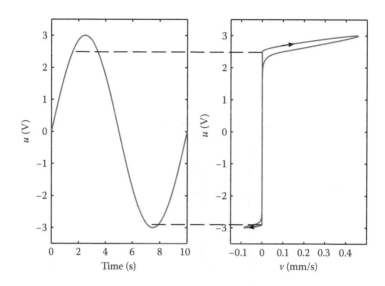

FIGURE 5.4
Identify static friction values using sine wave input.

voltage minus the friction. The correlations at very low speed and normal speed are measured. The asymmetric static friction values at which the motor starts moving, determined by injecting a sine wave function with low frequency and amplitude 3 V, are $f_{cp} = 2.5$ V, $f_{cn} = -2.9$ V, as seen from Figure 5.4.

Test inputs with bifrequency square waves and magnitude $u = \pm 5$ V and $u = \pm 3$ V are used to obtain the position response. The defining planes between the regions are taken at the velocity v_n, v_p obtained by applying $u = \pm 3$ V so the regions B and C safely encompass the nonlinear friction model. By removing the friction force from the test inputs, the two models $\{A_i, B_i\}$, $i = 1, 2$, are obtained and validated in Figure 5.5.

Denote the state $x = \begin{bmatrix} y & v \end{bmatrix}^T$. The piecewise affine model, after being transformed to discrete time, can be formally defined in the four convex subspaces:

$$x_{k+1} = \begin{cases} A_1 x_k + B_1(u_k - f_{cp}) & if\ v \geq v_p\ (\Omega_1) \\ A_1 x_k + B_1(u_k - f_{cn}) & if\ v \leq v_n\ (\Omega_2) \\ A_2 x_k + B_2(u_k - f_{cp}) & if\ 0 \leq v \leq v_p\ (\Omega_3) \\ A_2 x_k + B_2(u_k - f_{cn}) & if\ v_n \leq v \leq 0\ (\Omega_4) \end{cases} \tag{5.10}$$

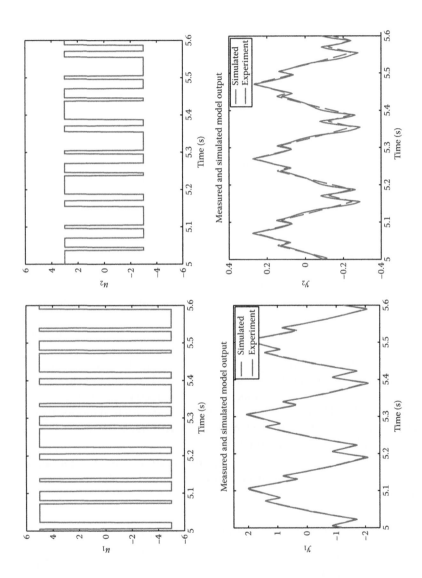

FIGURE 5.5
Model validation at two input ranges around $u_1 = 5$ (left) and $u_2 = 3$ (right).

5.2.2 Model Predictive Control for PWA Model

This section presents the integral model predictive design, including of state augmentation, MPC tracking formulation, and robust design.

Since the position tracking is emphasized, the new augmented system $(\bar{x}_k, \bar{u}_k, \bar{y}_k)$ is

$$
\begin{bmatrix} x_{k+1} \\ r_{k+1} \\ \theta_{k+1} \end{bmatrix} = \begin{bmatrix} A_i & B_i & 0 \\ 0 & 1 & 0 \\ C_i & -1 & 1 \end{bmatrix} \begin{bmatrix} x_k \\ r_k \\ \theta_k \end{bmatrix} + \begin{bmatrix} B_i \\ 1 \\ 0 \end{bmatrix} u_k
$$

$$
\bar{y}_k = [C_i \quad 0 \quad 0]\bar{x}_k \tag{5.11}
$$

where $r(k)$ is the reference position. Due to the motion system characteristics that this model description already contains an integrator, it is not necessary to use the Δu tracking formula here. Instead, integrating state θ_k guarantees zero-offset warranty.

An MPC optimal control scheme uses the system (5.11) to predict the output error ahead of time and uses current feedback errors to compensate any disturbance. The general form of MPC is stated as follows:

$$
V_N^o(\bar{x}_0, \bar{U}) = \min_{\bar{U}}. \bar{x}_N^T P_j \bar{x}_N + \sum_{k=0}^{N-1} \left(\bar{x}_k^T \bar{C}_j^T Q \bar{C}_j \bar{x}_k + \bar{u}_k^T R \bar{u}_k \right) \tag{5.12}
$$

$$
subj. to \quad \bar{x}_k \in \mathbb{X}, \bar{u}_k \in \mathbb{U} \quad k = 0, ..., N-1; \ \bar{x}_N \in X_{jf}
$$

$$
\bar{u}_N = K_j \bar{x}_k + d_j \qquad k \geq N
$$

where different prediction models in (5.10) are used if $\bar{x}_k \in \Omega_j$ ($j = 1, 2, 3, 4$). $\mathbb{X}, \mathbb{U}$ are the state and input constraint sets. After N control steps, the scheme expects the state to reside inside the terminal regions X_{jf}, which is also an control invariant set defined by a linear state feedback $\bar{u} = K_j \bar{x} + d_j$ (d_j: auxiliary input).

Stability analysis for piecewise affine systems using MPC has been analyzed carefully in [17] where the design of terminal cost P_j and terminal set X_{jf} is proposed. However, we focus more on the robust design aspect. Since regions A, D and B, C share the same linear dynamics, the MPC design only needs to consider two cases $\{A_i, B_i\}$ for $i = 1, 2$. The purpose is to keep the system state within a terminal constraint invariant set inside the dynamic $\{A_2, B_2\}$ (near the switching surface of the friction) by a robust gain K_2. Such a gain can be designed specifically for the dynamic P_2 with bounded disturbance assumption. The rest of this section describes a unique MPC component design.

5.2.2.1 Terminal Gain

Consider again the augmented dynamics from (5.10). Let $d_i = f_{cp}$ (or f_{cn}), and take the feedback input $\bar{u}^{fb} = K_i \bar{x}$ as $K_1 = K_{LQR}(Q, R)$ and K_2 such that the system

$$\bar{x}_{k+1} = \bar{A}_2 \bar{x}_k + \bar{B}_2 \left(\bar{u}_k^{fb} + w_k \right)$$
$$\bar{y}_k = \bar{C}_2 \bar{x}_k \tag{5.13}$$

is robust against the friction model mismatch w ($|w| \leq w^*$). This design can use one of many existing techniques in the literature to deal with input disturbance. For this application, a nonrecursive method for H_∞ [6] is applied to solve the related discrete-time Riccati equation (DARE). The obtained gain K_2 guarantees

$$\left\| T_{wy} \right\|_\infty \leq \gamma \tag{5.14}$$

where T_{wy} is the transfer function from w_k to y_k and γ is the infimum of the H_∞ design.

5.2.2.2 Terminal Cost

To guarantee the monotonous decrease of the cost function V_N^o inside the terminal set, the terminal cost P_i should satisfy

$$(A_i + B_i K_i)^T P_i (A_i + B_i K_i) - P_i \leq -Q - K_i^T R K_i \tag{5.15}$$

This condition can be solved efficiently using linear matrix inequalities (LMIs). Note that P_1 can be calculated easily by taking the equality in (5.15) and solving a discrete ARE.

5.2.2.3 Terminal Set

The common terminal set $X_f = X_{if}$ ($i = 1, 2$) is the maximal positively invariant set inside regions B and C and computed based on the gain K_2 and the system constraints. For an arbitrary set Z, define the operator $\Phi(Z) = \{\bar{x} | (A_2 + B_2 K_2)\bar{x} \in Z\}$. Let X_0 be the largest possible compact polyhedron such that

$$X_0 \subset \{\bar{x} | (x, u) \in \mathbb{X} \times \mathbb{U}\} \cap (\Omega_2 \cup \Omega_3)$$
$$X_k = \Phi(X_{k-1}) \cap X_{k-1}, \quad i = 1, 2, \dots \tag{5.16}$$

As proved in [17], this iterative procedure can be completed in finite steps and $X_f = \lim_{k \to \infty} X_k$. The minimal robust positively invariant set for this approach can be computed efficiently by the algorithm in [33].

5.3 From Parametric MPC to PID Gain Scheduling Controllers

The result from Section 5.2.2 holds when it is applied to either an *online* or *offline* MPC formulation. In this section, we particularly use parametric MPC (offline) to demonstrate the PID gain scheduling realization. The overall MPC controller design is implemented using multiparametric toolbox (MPT) [16].

Observe that Equation (5.12) minimizes a convex value function subject to a convex constraint set. The current state $\bar{x}_0$ can be considered a parameter for this problem. We have the following definition

Definition 5.1 (Critical Region)
A critical region is defined as the set of parameters $\bar{x}_0$ for which the same set of constraints is active at the optimum $(\bar{x}_0, \bar{U}^0(\bar{x}_0))$.

In other words, if the constraints in (5.12) are presented as $G\bar{U} \leq S\bar{x}_0 + W$ and A is an associated set of row index,

$$CR_A = \{\bar{x}_0 \in X_0 \mid G_i \bar{U}^0 = S_i \bar{x}_0 + W_i \text{ for all } i \in A\} \tag{5.17}$$

In [4,40], it is shown that these critical regions are a finite number of closed, nonoverlapped polyhedra, and they cover completely X_0. Since $\bar{U} = \{u_0, ..., u_{N-1}\}$, the same properties apply for u_0^0. Theorem 5.1 states the key result (see [5]).

Theorem 5.1 (Parametric Solution of MPC)
The optimal control law $u_0^0 = f(\bar{x}_0)$, $f : X_0 \mapsto U$, obtained as a solution of (5.12), is continuous and piecewise affine on the polyhedra

$$f(\bar{x}_0) = F^i \bar{x}_0 + g^i \quad if \ \bar{x}_0 \in CR^i, i = 1, ..., N^r, \tag{5.18}$$

where the polyhedral sets $CR^i \triangleq \{H^i \bar{x}_0 \leq k^i\}, i = 1, ..., N^r$ are a partition of the feasible set X_0.

Interestingly, the state feedback gain for $\bar{x}_0$ corresponds to velocity $v(k)$, position $x(k)$, and position integrating $\theta(k)$, which is similar to a PID structure (except the feedforward part because of the varying position reference and active constraints). Hence, this is similar to a gain scheduling feedforward PID where the decision variable is the current state $\bar{x}_0$, as shown in Figure 5.6.

The optimal input of MPC is applied for regions outside X_f. When $\bar{x}(k)$ reaches X_f, the system will be stabilized by the pure gain $F^0 = K$. Therefore, one practical way to design PID for constrained systems is to design a gain

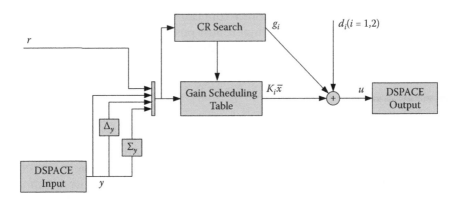

FIGURE 5.6
Proposed PID gain scheduling structure.

for its unconstrained region, which has been accomplished in Section 5.2.2, and applying these settings on the MPC formulation (5.12).

5.4 Simulation Study and Experiment Results

5.4.1 Simulation Studies

In this section, the previous theoretical development will be verified with a simulation study. The purpose is to obtain a set of parameters to apply in the real experiment.

First, MPC parameters are determined based on the nominal linear model $\{A_1, B_1\}$ without friction. The control horizon is chosen as $N = 5$. Weighting matrices Q, R are tuned by the guideline in [27] as $Q = diag\{10^4, 0.5, 10^4\}$, $R = 0.001$. This choice gives a relative good performance for large changes of set point (Figure 5.7). In fact, the response is fast since $u(k) = u_{max} = 10$ V only powers up the actuator to a velocity $v(k) = 180$ mm/s, smaller than $v_{max} = 400$ mm/s.

In another study, the robust effectiveness of the proposed control strategy is tested. To simulate the friction uncertainty, we assume that the USM has a similar linear model as the models identified from experiment data in Section 5.2.1, but a different friction form. The identified parameters are $f_{cp} = 2.5$ V, $f_{cn} = -2.9$ V, and

$$A_1 = \begin{bmatrix} 0.9968 & 6.289 \times 10^{-4} \\ -5.544 & 0.3623 \end{bmatrix}, \quad B_1 = \begin{bmatrix} 4.616 \times 10^{-3} \\ 3.493 \end{bmatrix},$$

$$A_2 = \begin{bmatrix} 0.9990 & 6.312 \times 10^{-4} \\ -1.658 & 0.3662 \end{bmatrix}, \quad B_2 = \begin{bmatrix} 2.033 \times 10^{-3} \\ 1.636 \end{bmatrix} \quad (5.19)$$

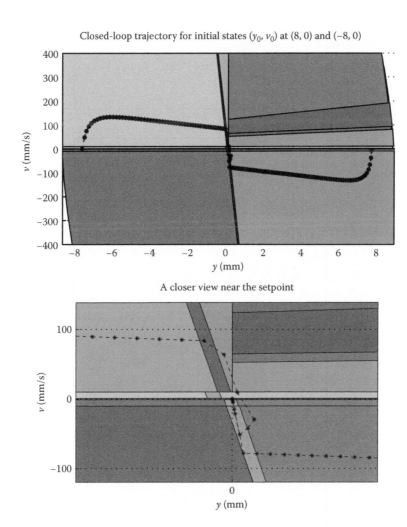

FIGURE 5.7
Performance of a normal MPC controller tuned by [27] on the outer model $\{A_1, B_1\}$.

while the assumed real parameters are $f_{cp} = 2.4$ V, $f_{cn} = -2.9$ V, A_1, B_1 unchanged, and

$$
A_2 = \begin{bmatrix} 0.9990 & 6.312 \times 10^{-4} \\ -1.658 & 0.4000 \end{bmatrix}, \quad B_2 = \begin{bmatrix} 2.033 \times 10^{-3} \\ 1.636 \end{bmatrix} \tag{5.20}
$$

An increase in $A_2(2, 2)$ represents a steeper negative slope of f_s (Figure 5.3).

Two MPC schemes are compared: LQR MPC and the proposed controller with the additional robust design. Both MPCs are designed with the tuned parameters above. Figure 5.8 shows the tracking errors, velocities, and inputs

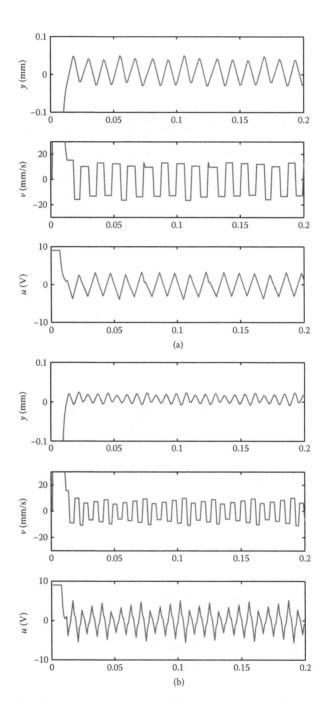

FIGURE 5.8
Simulated output errors with friction model mismatch for (a) LQR MPC design and (b) proposed
robust MPC.

after a step change in the reference. It can be seen that when friction mismatch presents, the output error shows an oscillation around the set point. Through the H_∞ design, the oscillation magnitude is substantially reduced from 0.03 mm to 0.014 mm. Hence, the effect of mitigating the friction model error is demonstrated by the proposed method.

5.4.2 Experiment

In this section, real-time experiments are carried out on an ultrasonic drive system. The setup uses the PI M-663 with velocity limit 400 mm/s and travel range 19 mm. The dSPACE control development and rapid prototyping system, in particular, the DS1104 board, is used, which integrates the whole development cycle seamlessly into a single environment. MATLAB/Simulink can be directly used in dSPACE. The sampling period for this test is chosen as 1 ms.

The commonly used relay PID tuning is chosen to compare with the proposed method. In this experiment, the system model at $u = 3$ V is used for relay tuning to obtain PID gain

$$K = \begin{bmatrix} 34.96 & 0.09674 & 1545.5 \end{bmatrix} \tag{5.21}$$

While it may not offer the best PID tuning in all situations, relay PID exhibits a large integrating factor to overcome the friction, thus achieving a fast rising time and zero offset. These characteristics can be used to evaluate the performance of the MPC method.

Parametric programming is used to solve the MPC problem for PWA models. The obtained controller is implemented under a lookup table form with 23 regions where only matrix multiplication and comparison are performed.

$$u_k = K_i \bar{x}_k + d_i \text{ if } \bar{x}_k \in C R_i \quad i = 1, 2, ..., 23 \tag{5.22}$$

This feasible form is no more complicated than the traditional PID plus nonlinear compensations.

Additionally, in order to remove the error oscillation observed in the simulation studies, a dead band with small $\epsilon > 0$ is imposed for the input.

$$u_k = \begin{cases} 0 & if \quad |v| \le \epsilon \\ u_{MPC} & if \quad |v| > \epsilon \end{cases} \tag{5.23}$$

The dead band can be implemented as a precondition prior to evaluating (5.22).

The tracking response of a square-wave trajectory ($f = 1$ Hz, $A = 1$ mm) is shown in Figure 5.9. In Figure 5.9(a) the tracking output shows that relay PID creates an overshoot created about 0.5 mm, which is undesirable. The proposed method can achieve a rising time as fast as the large-integrator PID, and produce no overshoot. Hence, improved steady-state tracking is shown.

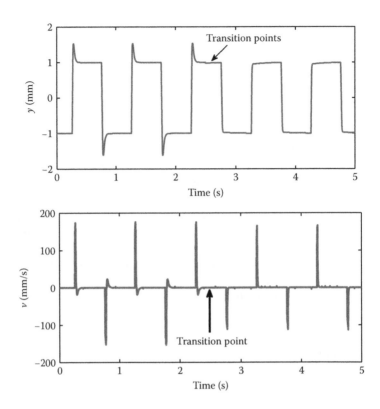

FIGURE 5.9
Experiment results when comparing relay PID (before) and the proposed method (after).

Figure 5.9(b) shows a smooth decreasing of velocity toward zero, implying that the friction uncertainty is around the stationary point.

5.5 Concluding Remarks

A robust MPC method has been developed for compensation of friction arising in linear USM. The objective of the control scheme is to achieve good static tracking performance in the presence of uncertainties in the friction model. This is obtained by incorporating linear friction inside the hybrid plant model and designing a robust terminal gain for MPC. Simulation and experimental results have shown that the proposed compensation technique can overcome the limitations of the relay PID tuning while attaining a simple real-time implementation.

References

1. D. Angeli, A. Casavola, and E. Mosca. Predictive PI-control of linear plants under positional and incremental input saturations. *Automatica*, 36(10), 1505–1516, 2000.
2. B. Armstrong and B. Amin. PID control in the presence of static friction: A comparison of algebraic and describing function analysis. *Automatica*, 32(5), 679–692, 1996.
3. V.L. Bageshwar and F. Borrelli. On a property of a class of offset-free model predictive controllers. *IEEE Transactions on Automatic Control*, 54(3), 663–669, 2009.
4. M. Baotic. An efficient algorithm for multiparametric quadratic programming. Technical report. ETH, April 2002.
5. A. Bemporad, M. Morari, V. Dua, and E. N. Pistikopoulos. The explicit linear quadratic regulator for constrained systems. *Automatica*, 38(1), 3–20, 2002.
6. M. B. Chen, A. Saberi, and Y. Shamash. A non-recursive method for solving the general discrete-time Riccati equations related to the H control problem. *International Journal of Robust and Nonlinear Control*, 4(4), 503–519, 1994.
7. P. Dupont, V. Hayward, B. Armstrong, and F. Altpeter. Single state elastoplastic friction models. *IEEE Transactions on Automatic Control*, 47(5), 787–792, 2002.
8. J. Brian Froisy. Model predictive control—Building a bridge between theory and practice. *Computers and Chemical Engineering*, 30, 1426–1435, 2006.
9. H. Fujioka, C.-Y. Kao, S. Almér, and U. Jönsson. Robust tracking with performance for PWM systems. *Automatica*, 45(8), 1808–1818, 2009.
10. C. E. Garcia, D. M. Prett, and M. Morari. Model predictive control: Theory and practice—A survey. *Automatica*, 25(3), 335–348, 1989.
11. A. Grancharova, T. A. Johansen, and J. Kocijan. Explicit model predictive control of gas and liquid separation plant via orthogonal search tree partitioning. *Computers and Chemical Engineering*, 28(12), 2481–2491, 2004.
12. V. Hayward, B. S. R. Armstrong, F. Altpeter, and P. E. Dupont. Discrete-time elastoplastic friction estimation. *IEEE Transactions on Control Systems Technology*, 17(3), 688–696, 2009.
13. M. Herceg, M. Kvasnica, and M. Fikar. Minimum-time predictive control of a servo engine with deadzone. *Control Engineering Practice*, 17(11), 1349–1357, 2009.
14. Z.-P. Jiang and I. Marcels. Robust nonlinear integral control. *IEEE Transactions on Automatic Control*, 46(8), 1336–1342, 2001.
15. E. C. Kerrigan and J. M. Maciejowski. Robustly stable feedback min-max model predictive control. In *Proceedings of the 2003 American Control Conference*, June 2003, vol. 4, pp. 3490–3495.
16. M. Kvasnica, P. Grieder, and M. Baoti. Multi-parametric toolbox (MPT). Swiss Federal Institute of Technology, 2004.
17. M. Lazar, W. P. M. H. Heemels, S. Weiland, and A. Bemporad. Stabilizing model predictive control of hybrid systems. *IEEE Transactions on Automatic Control*, 51(11), 1813–1818, 2006.
18. C. Lvass, M. M. Seron, and G. C. Goodwin. Robust output-feedback MPC with integral action. *IEEE Transactions on Automatic Control*, 55(7), 1531–1543, 2010.

19. U. Maeder, F. Borrelli, and M. Morari. Linear offset-free model predictive control. *Automatica*, 45(10), 2214–2222, 2009.

20. U. Maeder and M. Morari. Offset-free reference tracking with model predictive control. *Automatica*, 46(9), 1469–1476, 2010.

21. L. Magni, G. De Nicolao, R. Scattolini, and F. Allgwer. Robust model predictive control for nonlinear discrete-time systems. *International Journal of Robust and Nonlinear Control*, 13(3–4), 229–246, 2003.

22. P. Marquis and J. P. Broustail. Smoc, a bridge between state space and model predictive controllers: Application to the automation of a hydrotreating unit. In *IFAC Workshop on Model Based Process Control*, 1998.

23. D. Q. Mayne, J. B. Rawlings, C. V. Rao, and P. O. M. Scokaert. Constrained model predictive control: Stability and optimality. *Automatica*, 36, 789–814, 2000.

24. M. Morari and U. Maeder. Nonlinear offset-free model predictive control. *Automatica*, 48(9), 2059–2067, 2012.

25. K. R. Muske and T. A. Badgwell. Disturbance modeling for offset-free linear model predictive control. *Journal of Process Control*, 12(5), 617–632, 2002.

26. K. R. Muske and J. B. Rawlings. Model predictive control with linear models. *AIChE Journal*, 39(2), 262–287, 1993.

27. M. H. T. Nguyen, K. K. Tan, and S. Huang. Enhanced predictive ratio control of interacting systems. *Journal of Process Control*, 21(7), 1115–1125, 2011.

28. G. Pannocchia and J. B. Rawlings. Disturbance models for offset-free model-predictive control. *AIChE Journal*, 49(2), 426–437, 2003.

29. G. Pannocchia, J. B. Rawlings, and S. J. Wright. Fast, large-scale model predictive control by partial enumeration. *Automatica*, 43(5), 852–860, 2007.

30. U. Parlitz, A. Hornstein, D. Engster, F. Al-Bender, V. Lampaert, T. Tjahjowidodo, S. D. Fassois, D. Rizos, C. X. Wong, K. Worden, and G. Manson. Identification of pre-sliding friction dynamics. *Chaos*, 14(2), 420–430, 2004.

31. Y.-F. Peng and C.-M. Lin. Intelligent motion control of linear ultrasonic motor with h-infinite tracking performance. *IET Control Theory Applications*, 1(1), 9–17, 2007.

32. G. V. Raffo, M. G. Ortega, and F. R. Rubio. An integral predictive/nonlinear control structure for a quadrotor helicopter. *Automatica*, 46(1), 29–39, 2010.

33. S. V. Rakovic, E. C. Kerrigan, K. I. Kouramas, and D. Q. Mayne. Invariant approximations of the minimal robust positively invariant set. *IEEE Transactions on Automatic Control*, 50(3), 406–410, 2005.

34. J. B. Rawlings and D. Q. Mayne. *Model predictive control: Theory and design*. Madison, WI: Nob Hill Publishing, 2009.

35. J. Richalet, A. Rault, J. L. Testud, and J. Papon. Model predictive heuristic control: Applications to industrial processes. *Automatica*, 14(5), 413–428, 1978.

36. V. Sakizlis, N. M. P. Kakalis, V. Dua, J. D. Perkins, and E. N. Pistikopoulos. Design of robust model-based controllers via parametric programming. *Automatica*, 40(2), 189–201, 2004.

37. P. O. M. Scokaert and D. Q. Mayne. Min-max feedback model predictive control for constrained linear systems. *IEEE Transactions on Automatic Control*, 43(8), 1136–1142, 1998.

38. S. Seshagiri and H. K. Khalil. Robust output feedback regulation of minimum-phase nonlinear systems using conditional integrators. *Automatica*, 41(1), 43–54, 2005.

39. K. K. Tan, S. N. Huang, and T. H. Lee. Development of a GPC-based PID controller for unstable systems with deadtime. *ISA Transactions*, 39(1), 57–70, 2000.
40. P. Tondel, T. A. Johansen, and A. Bemporad. An algorithm for multi-parametric quadratic programming and explicit MPC solutions. *Automatica*, 39(3), 489–497, 2003.
41. M. Vasak, M. Baotic, I. Petrovic, and N. Peric. Hybrid theory-based time-optimal control of an electronic throttle. *IEEE Transactions on Industrial Electronics*, 54(3), 1483–1494, 2007.
42. C. Yousfi and R. Tournier. Steady state optimization inside model predictive control. In *American Control Conference*, June 1991, pp. 1866–1870.
43. Y. Zhang, L.-S. Shieh, and A. C. Dunn. PID controller design for disturbed multivariable systems. *IEE Proceedings Control Theory and Applications*, 151(5), 567–576, 2004.
44. K. Zhou and J. C. Doyle. *Essentials of robust control*. Englewood Cliffs, NJ: Prentice-Hall, 1998.

6

Modeling and Control of Air Bearing Stages

In conventional motion stage designs, multiple motors and coupling mechanisms (like gears and link mechanisms) are used to implement multi-DOF (degrees of freedom) movements. However, such kinds of designs will incur friction, which degrades the control precision. Motivated by this phenomenon, air bearing stages have been suggested to overcome friction. An air bearing system is based on a pressurized thin film of air to support an applied normal load [1]. For over 30 years, the industry has found it can move heavy, cumbersome loads and equipment easier and faster by using frictionless air. Moving heavy loads on air is a clean, quiet, and safe method that will not damage floors. Comparing the air bearings with other types of bearings (including rolling bearings and fluid bearings), air bearing has low viscosity and offers lower friction and smaller sensitivity to temperature variation [2].

Various air bearing control systems can be found in the literature (see Ezenekwe and Lee [3], Fan et al. [2], Zhang et al. [4], Schwartz and Hall [5], Sescu et al. [6]). However, these results are based on open-loop control. It is found that due to very low damping, the control of the air bearing system is challenging work. For the closed-loop control, adopting air bearing can greatly help a multi-DOF system to improve its control performance. A multi-DOF spherical air bearing system is found in [7], and two different three-DOF precision positioning systems using direct-drive motors/actuators and air bearing are illustrated in [8] and [9].

The purpose of this chapter is to present the control designs for air bearing precise stages. Unlike general mechanical motion control systems, the air bearing system involves a floating carriage, and it is thus very sensitive to the system noise and vibration; this is a challenging issue in the control design of an air bearing stage. To solve such problems, we first consider the control problem of a linear air bearing stage, and eddy current braking is introduced. Only proportional-integral (PI) control is used in the system for avoiding possible vibration. Subsequently, we consider the control problem of a multi-DOF spherical air bearing stage. An adaptive noise filter and a controller for angular positioning are proposed to achieve high performance. Finally, experimental results are given to show the effectiveness of the proposed control algorithm.

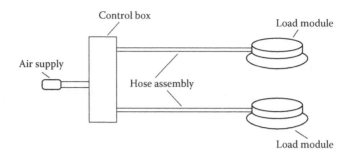

FIGURE 6.1
A typical air bearing system.

6.1 Problem Statements

Typical air bearing systems are similar to the diagram shown here (see Figure 6.1). This is a two-station air bearing system consisting of two load modules and a control unit. The control box controls the airflow to each individual air bearing. Once correct airflow has been achieved, the result is an air bearing system that floats across the floor in a free and even manner. Air bearings can easily solve machinery and load moving problems: float them across the floor using air-powered air bearings, and with pinpoint accuracy and load positioning. However, due to the very low damping, the control of the air bearing system is challenging work. To solve this issue, new control methods should be introduced. In the following, we will discuss two ways for two different types of air bearing stages.

6.2 Control of Linear Air Bearing Stage

A linear air bearing stage is a system, usually incorporating air bearing to form a floating action above the linear bar surface. In this chapter, the film of air has a constant thickness, for we maintain a constant pressure and the carriage and load are not changed during our experimental tests. However, due to the very low damping, the control of the air bearing system is challenging work. To solve this issue, an eddy current damper can be introduced into the system.

Eddy currents are created in a conductor in a magnetic field. They are induced either by the movement of the conductor in the static field or by changing the strength of the magnetic field, initiating motional and transformer electromotive forces (EMFs), respectively. Since the created eddy currents

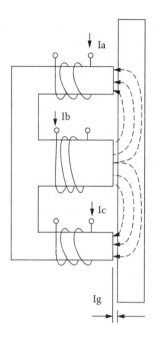

FIGURE 6.2
Linear eddy current brake.

produce a force that is proportional to the velocity of the conductor, the moving magnet and conductor behave like a viscous damper. Figure 6.2 shows our design of an eddy current system. The integration of the air bearing system and eddy current braking is shown in Figure 6.3. The system consists of air pressure, linear actuator, eddy current damping, and controller.

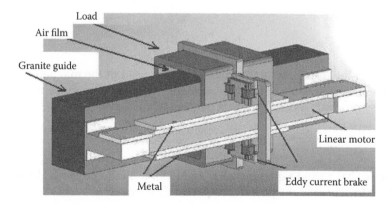

FIGURE 6.3
Linear air bearing system with eddy current brake.

1. **Air pressure system:** Air pressure provides gas to support the carriage to freely move along the linear granite guide. It consists of compressor, filter, and load modules.
2. **Linear actuator:** The linear actuator is used to drive the carriage supported by air bearings. One linear optical encoder is mounted to the linear guide and provides positioning information.
3. **Eddy current damping:** The eddy current damping can help the air bearing stage to generate a necessary damping when the carriage is required to be controlled.
4. **Controller:** The controller is designed to provide the strategy for both the linear actuator and the eddy current. In the following, we focus on the control algorithm for the air bearing stage.

The mathematical models of the linear air bearing system can be derived from the Lorentz force law and Joule effect [10,11].

The flux is related to the current by the magneto motive force (6.1) relationship:

$$\Phi = \frac{Ni}{\mathcal{R}} \tag{6.1}$$

where $\mathcal{R}$ is the reluctance of the magnetic circuit, N is the number of turns in the coil, and i is the coil current. Assuming the reluctance of the two air gaps is much larger than that of the magnetic material in the pole piece, $\mathcal{R}$ can be approximated by

$$\mathcal{R} = \frac{l_g}{\mu_0 S} \tag{6.2}$$

where l_g is the length of one air gap, μ_0 is the permeability of free space, and S is the area of the pole piece normal to the flux direction.

The current density J, induced in the conducting sheet due to the motional EMF, is

$$J = \frac{1}{\rho}(\dot{x} \times B) \tag{6.3}$$

where ρ is the resistivity of the metal, $\dot{x}$ is the speed of the magnetic system (it is the time derivative of the displacement x of the system movement), $\times$ is the cross-product, and B is the flux density, that is given by

$$B = \frac{\Phi}{S} = \frac{Ni}{S\mathcal{R}} = \frac{\mu_0 Ni}{l_g} \tag{6.4}$$

where we have used Equations (6.1) and (6.2). If all the power is dissipated by the eddy current, the power dissipation P_d is given by

$$P_d = \rho J^2 \times Volume = \frac{1}{\rho}\dot{x}^2 B^2 Sd \tag{6.5}$$

where d is the metal thickness. The braking force T_b generated by P_d can be expressed as

$$T_b = \frac{P_d}{\dot{x}} = \frac{1}{\rho} Sd \left(\frac{\mu_0 N}{l_g}\right)^2 i^2 \dot{x} \tag{6.6}$$

where we have used Equation (6.4).

From Newton's second law, the air bearing system with the eddy current brake can be described by a 1-DOF mathematical equation

$$m\ddot{x} = u(t) - T_b - f \tag{6.7}$$

where $u(t)$ is the time-varying motor control force, $x(t)$ is the position of the air bearing with eddy current brake, T_b is the eddy current braking force, m is the effective mass, and f is the frictional force. Here, the eddy current brake is given by $T_b = Ti^2\dot{x}$, where T is the braking force constant $T = \sigma Sd(\frac{\mu_0 N}{l_g})^2$. Since the air bearing system with linear actuator will experience some friction, the force f is assumed to be $f = D\dot{x}$.

6.3 Controller Design for the System

The dynamical modeling of the designed air bearing system with damper has been stated in the last section. In this section, we design a controller to achieve precision tracking control.

Consider the system (6.7) and rewrite it as

$$\ddot{x} = \frac{1}{m}u - \frac{T}{m}i^2\dot{x} - \frac{D}{m}\dot{x} \tag{6.8}$$

The control problem is to design a controller for the system (6.7) such that the position x follows a trajectory x_d closely as possible.

Define $e = x_d - x$ and $e_v = \dot{x}_d - \dot{x}$. The above equation produces

$$\left.\begin{aligned} \dot{e} &= e_v \\ \dot{e}_v &= \ddot{x}_d - \frac{D}{m}e_v + \frac{D}{m}\dot{x}_d - \frac{1}{m}u - \frac{T}{m}i^2 e_v \\ &\quad + \frac{T}{m}i^2\dot{x}_d \end{aligned}\right\} \tag{6.9}$$

Remark 1: It should be noted that there are two variables u and i involved in the control inputs. Obviously, this is a nonlinear control problem. Unlike general mechanical systems, the air bearing system is a floating carriage, and it is thus very sensitive to the system noise and vibration. This is because in normal situations the floating carriage is in equilibrium, and once noise enters to the air bearing system, the floating carriage may lose its balance, thereby causing a chattering phenomenon. The traditional PID controller involves three separate values: the proportional, integral, and derivative values. Due to the derivative value determining the reaction based on the rate at which the error has been changing, we will not use this term in the controller to avoid any possible vibration for the air bearing system. In what follows, we will propose a control scheme dealing with the air bearing system without requiring velocity information.

The proposed control algorithm is given by

$$\left.\begin{aligned}
u &= Ke + Ti^2\dot{x}_d + m\ddot{x}_d + D\dot{x}_d \\
i &= \qquad K_i\sqrt{|e|}
\end{aligned}\right\} \tag{6.10}$$

where K is the feedback control gain of the control input u, K_i is the control gain of the control input i, and T, m, and D are the estimated parameters.

Substituting the control law (6.10) into the above equation yields

$$\dot{e} = e_v \tag{6.11}$$

$$\dot{e}_v = -\frac{K}{m}e - \frac{D}{m}e_v - \frac{T}{m}K_i^2|e|e_v \tag{6.12}$$

Introducing the variable $X = [e \ e_v]^T$, we have

$$\dot{X} = AX - b\frac{T}{m}K_i^2|e|e_v \tag{6.13}$$

where

$$A = \begin{bmatrix} 0 & 1 \\ -\frac{K}{m} & -\frac{D}{m} \end{bmatrix}, \quad b = \begin{bmatrix} 0 \\ 1 \end{bmatrix} \tag{6.14}$$

Note that A is a stable matrix. Let α, λ be the positive constants that satisfy $\|e^{A(t-\tau)}\| \le \alpha e^{-\lambda(t-\tau)}$.

Assumption 6.1

The desired trajectories x_d, $\dot{x}_d$, and $\ddot{x}_d$ are bounded, i.e., $|x_d| \le x_{dM}$, $|\dot{x}_d| \le x_{dM}^{(1)}$, $|\ddot{x}_d| \le x_{dM}^{(2)}$ with the constants x_{dM}, $x_{2dM}^{(1)}$, and $x_{dM}^{(2)}$.

Theorem 6.1

Consider the system (6.7). If the control law (6.10) is applied to the system, then for $K > 0$, $K_i > 0$:

1. *$\lim_{t \to \infty} |e_v| = 0$, i.e., for given small number ϵ, we have $|e_v| \le \epsilon$ for $t \ge t_1 > 0$.*

2. *Furthermore, when $\lambda > \frac{\alpha \epsilon T K_i^2 \|b\|}{m}$, then the tracking error e converges to zero.*

Proof

Define the Lyapunov function:

$$V = \frac{K}{2m}e^2 + \frac{1}{2}e_v^2 \qquad (6.15)$$

Its time derivative is given by

$$\dot{V} = \frac{K}{m}ee_v + e_v\left(-\frac{K}{m}e - \frac{D}{m}e_v - \frac{T}{m}K_i^2|e|e_v\right)$$

$$= -\frac{D}{m}e_v^2 - \frac{T}{m}K_i^2|e|e_v^2 \qquad (6.16)$$

This implies that e and e_v are bounded. From (6.11) and (6.12), this also implies that $\dot{e}_v$ is bounded. Moreover, from (6.16), we obtain that $\dot{V} \leq -\frac{D}{m}e_v^2$. This implies that

$$\frac{D}{m}\int_0^t e_v^2 d\tau \leq V(0) - V(t) \qquad (6.17)$$

By using Barbalat's lemma, it is shown that $\lim_{t\to\infty}|e_v| = 0$ and the proof of (1) is completed.

Let us now see the composite equation (6.13). The solution of Equation (6.13) is given by

$$X = e^{A(t-t_0)}X(t_0) - \int_{t_0}^t e^{A(t-\tau)}\frac{bTK_i^2}{m}|e|e_v d\tau \qquad (6.18)$$

Because A is a stable matrix, there exist positive constants α, λ such that $||e^{At}|| \leq \alpha e^{-\lambda t}$. Using this result, we have

$$||X|| \leq \alpha e^{-\lambda(t-t_0)}||X(t_0)|| + \int_{t_0}^t \alpha e^{-\lambda(t-\tau)} \left|\left|\frac{bTK_i^2}{m}\right|\right| \ ||e||e_v| d\tau$$

Since $|e| \leq ||X||$, this implies that

$$|e| \leq \alpha e^{-\lambda(t-t_0)}||X(t_0)|| + \int_{t_0}^t \alpha e^{-\lambda(t-\tau)} \left|\left|\frac{bTK_i^2}{m}\right|\right| \ ||e||e_v| d\tau$$

From the conclusion of that, $\lim_{t\to\infty}|e_v| = 0$; this implies that for a given small number ϵ, we have $|e_v| \leq \epsilon$ for $t \geq t_1 > 0$. Thus, we have

$$|e| \leq \alpha e^{-\lambda(t-t_1)}||X(t_1)|| + \int_{t_1}^t \alpha\epsilon e^{-\lambda(t-\tau)} \left|\left|\frac{bTK_i^2}{m}\right|\right| \ ||e| d\tau$$

for $t \geq t_1 > 0$. Utilizing Gronwall's lemma, we have

$$|e| \leq \alpha e^{-\lambda(t-t_1)} ||X(t_1)|| + \frac{\alpha^2 \epsilon T K_i^2 ||b||}{m} ||X(t_1)|| e^{\lambda t_1} e^{-(\lambda - \frac{\alpha \epsilon T K_i^2 ||b||}{m})t} \int_{t_1}^{t} e^{-\frac{\alpha \epsilon T K_i^2 ||b||}{m} \tau} d\tau$$

$$\leq \alpha e^{-\lambda(t-t_1)} ||X(t_1)|| + \alpha ||X(t_1)|| e^{\lambda t_1} e^{-(\lambda - \frac{\alpha \epsilon T K_i^2 ||b||}{m})t}$$

for $t \geq t_1$. This shows that $\lim_{t \to \infty} |e| = 0$ if $\lambda > \frac{\alpha \epsilon T K_i^2 ||b||}{m}$. The proof of (2) is completed.

Remark 2: Our stability analysis involves the estimated parameters T, m, D and design constants K, K_i. The condition $\lambda > \frac{\alpha \epsilon T K_i^2 ||b||}{m}$ in Theorem 6.1 can be met easily, for a small number ϵ can be given.

Remark 3: In a practical system, it is quite difficult to know the exact values of the model parameters. In this case, we have to study the robustness of the controlled system. Suppose that the actual values of D, m, T (denoted by D^*, m^*, T^*) are different from their identified values D, m, T, respectively. It is reasonable to assume that we know the ranges of the actual values D^*, m^*, T^* whose elements lie in known bounded sets such that $D_m \leq D^* \leq D_M$, $m_m \leq m^* \leq m_M$, $T_m \leq T^* \leq T_M$. Since the interval $[D_m, D_M]$ is with center $O_D = (D_m + D_M)/2$ and radius $R_D = (D_M - D_m)/2$, we have $|D^* - D| \leq R_D + |D - O_D| = \frac{D_M - D_m + |2D - D_m - D_M|}{2}$. Similarly, we have $|m^* - m| \leq \frac{m_M - m_m + |2m - m_m - m_M|}{2}$, $|T^* - T| \leq \frac{T_M - T_m + |2T - T_m - T_M|}{2}$. In what follows, we discuss how the errors (between D^* and D, m^* and m, or T^* and T) will affect the proposed controller under the control form (6.10).

Recalling the system (6.9) under the actual values denoted by D^*, m^*, T^*, we have

$$\left. \begin{array}{l} \dot{e} = e_v \\ \dot{e}_v = \ddot{x}_d - \frac{D^*}{m^*} e_v + \frac{D^*}{m^*} \dot{x}_d - \frac{1}{m^*} u - \frac{T^*}{m^*} i^2 e_v \\ \qquad + \frac{T^*}{m^*} i^2 \dot{x}_d \end{array} \right\} \tag{6.19}$$

The control law (6.10) with the estimates D, m, T is applied to the system, and we establish the following theorem:

Theorem 6.2

Consider the system (6.19). If the control law (6.10) is applied to the system, then for $K > 0$, $K_i > 0$:

1. e_v *is uniformly bounded and*

$$|e_v| \leq B_v = max \left\{ \frac{m_M d_1}{D_m}, \frac{m_M d_2}{T_m K_i^2} \right\} \tag{6.20}$$

where

$$d_1 = \frac{m_M - m_m + |2m - m_m - m_M|}{2m_m} x_{dM}^{(2)}$$

$$+ \frac{D_M - D_m + |2D - D_m - D_M|}{2m_m} x_{dM}^{(1)} \qquad (6.21)$$

$$d_2 = \frac{T_M - T_m + |2T - T_m - T_M|}{2m_m} K_i^2 x_{dM}^{(1)} \qquad (6.22)$$

2. *Furthermore, when*

$$\lambda > \alpha \, \|b\| \left(\frac{T_M K_i^2}{m_m} B_v + d_2 + d_3 \right) \qquad (6.23)$$

where

$$d_3 = \frac{K(m_M - m_m + |2m - m_m - m_M|)}{2mm_m} \qquad (6.24)$$

then the tracking error e remains bounded and

$$\lim_{t \to \infty} |e| \le \frac{\alpha(d_1 + d_4 B_v)\|b\|}{\lambda - \alpha \, \|b\| \left(\frac{T_M K_i^2}{m_m} B_v + d_2 + d_3 \right)} \qquad (6.25)$$

where

$$d_4 = \frac{D}{m} \frac{m_M - m_m + |2m - m_m - m_M|}{2m_m}$$

$$+ \frac{D_M - D_m + |2D - D_m - D_M|}{2m_m} \qquad (6.26)$$

Proof

Substituting the control law (6.10) into the system (6.19) yields

$$\dot{e} = e_v \qquad (6.27)$$

$$\dot{e}_v = \bar{m}\ddot{x}_d + \bar{D}\dot{x}_d + \bar{T}K_i^2|e|\dot{x}_d - \frac{K}{m^*}e - \frac{D^*}{m^*}e_v - \frac{T^*}{m^*}K_i^2|e|e_v \qquad (6.28)$$

where $\bar{m} = \frac{m^* - m}{m^*}, \bar{D} = \frac{D^* - D}{m^*}, \bar{T} = \frac{T^* - T}{m^*}$. Define the Lyapunov function $V = \frac{K}{2m^*}e^2 + \frac{1}{2}e_v^2$. Its time derivative is given by

$$\dot{V} = \frac{K}{m^*}ee_v + e_v \left(-\frac{K}{m^*}e - \frac{D^*}{m^*}e_v - \frac{T^*}{m^*}K_i^2|e|e_v \right) + e_v\left(\bar{m}\ddot{x}_d + \bar{D}\dot{x}_d + \bar{T}K_i^2|e|\dot{x}_d\right)$$

$$\le -|e_v| \left(\frac{D_m|e_v|}{m_M} - d_1 \right) - |e||e_v| \left(\frac{T_m K_i^2|e_v|}{m_M} - d_2 \right) \qquad (6.29)$$

This shows that $\dot{V} < 0$ if

$$|e_v| > max\left\{\frac{m_M d_1}{D_m}, \frac{m_M d_2}{T_m K_i^2}\right\} \tag{6.30}$$

This implies that e_v is uniformly bounded and $|e_v| \leq B_v = max\{\frac{m_M d_1}{D_m}, \frac{m_M d_2}{T_m K_i^2}\}$ for $t \geq t_2 > 0$. The proof of (1) is completed.

As in Theorem 6.1, we introduce the variable $X = [e \quad e_v]^T$ and rewrite Equations (6.27) and (6.28) as

$$\dot{X} = AX + b\left[\frac{K(m^* - m)}{mm^*}e + \frac{Dm^* - D^*m}{mm^*}e_v\right.$$
$$\left. - \frac{T^* K_i^2}{m^*}|e|e_v + \bar{m}\ddot{x}_d + \bar{D}\dot{x}_d + \bar{T}K_i^2|e|\dot{x}_d\right] \tag{6.31}$$

The solution of the above equation is given by

$$X = e^{A(t-t_2)}X(t_2) + \int_{t_2}^{t} e^{A(t-\tau)}b\left[\frac{K(m^* - m)}{mm^*}e + \frac{Dm^* - D^*m}{mm^*}e_v\right.$$
$$\left. - \frac{T^* K_i^2}{m^*}|e|e_v + \bar{m}\ddot{x}_d + \bar{D}\dot{x}_d + \bar{T}K_i^2|e|\dot{x}_d\right]d\tau \tag{6.32}$$

Since A is a stable matrix, there exists the inequality $||e^{At}|| \leq \alpha e^{-\lambda t}$ as in Theorem 6.1. Using this result, it follows that

$$||X|| \leq \alpha e^{-\lambda(t-t_2)}||X(t_2)|| + \int_{t_2}^{t}\alpha e^{-\lambda(t-\tau)}\,||b||$$
$$\times \left[\frac{K|m^* - m|}{mm^*}|e| + \left(\frac{D|m^* - m|}{mm^*} + \frac{|D - D^*|}{m^*}\right)|e_v|\right.$$
$$\left. + \frac{T^* K_i^2}{m^*}|e||e_v| + d_1 + d_2|e|\right]d\tau$$
$$\leq \alpha e^{-\lambda(t-t_2)}||X(t_2)|| + \frac{\alpha(d_1 + d_4 B_v)||b||}{\lambda}$$
$$+ \int_{t_2}^{t} e^{-\lambda(t-\tau)}\alpha\,||b||\left(\frac{T_M K_i^2}{m_m}B_v + d_2 + d_3\right)|e|d\tau$$

where we have used the properties as shown in *Remark 3*. Using Gronwall's lemma, we have

$$|e| \le \alpha \left(||X(t_2)|| + \frac{d_1 + d_4 B_v}{\lambda} ||b|| \right) e^{-(\lambda - c_1)(t - t_2)}$$

$$+ \alpha(d_1 + d_4 B_v) ||b|| \int_{t_2}^{t} e^{-(\lambda - c_1)(t - \tau)} d\tau$$

$$\le \alpha \left(||X(t_2)|| + \frac{d_1 + d_4 B_v}{\lambda} ||b|| \right) e^{-(\lambda - c_1)(t - t_2)} + \frac{\alpha(d_1 + d_4 B_v) ||b||}{\lambda - c_1}, \ t \ge t_2$$

where $c_1 = \alpha ||b|| (\frac{T_M K_l^2}{m_m} B_v + d_2 + d_3)$. If $\lambda > c_1$, the error e is bounded and $\lim_{t \to \infty} |e| \le \frac{\alpha(d_1 + d_4 B_v) ||b||}{\lambda - c_1}$. The proof of (2) is completed.

6.4 Experimental Results

In this section, we present the experimental tests using the proposed control algorithm. The motion system as shown in Figure 6.4 is a single-axis translation stage derived from an air bearing stage, with a load flying on a

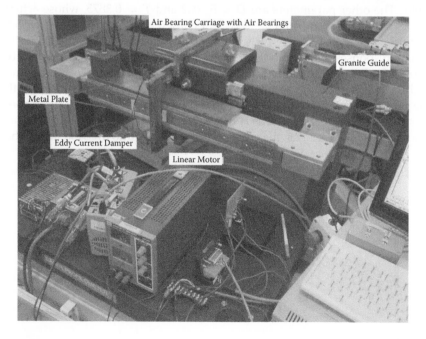

FIGURE 6.4
Experimental set-up.

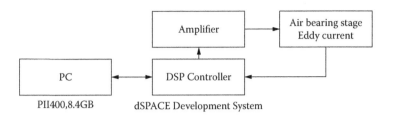

FIGURE 6.5
Functional block of control system.

granite guide and using pressure/vacuum air bearing technology. There are two control units in the system: linear motor control and eddy current control. A linear servomotor actuates the air bearing stage along a granite guide. A high-resolution linear optical encoder comes standard with resolutions down to 1 μm. The eddy current control unit is a circuit supported by a 12 V power. Both linear motor and eddy current units are controlled by the dSPACE control card, which is based on a PC with dSPACE card 1103. The functional block diagram is shown in Figure 6.5. The sampling period for our test is chosen as 0.0004 sec.

In Tan et al. [12], we have obtained the model parameters that can be used in the controller design. The mass can be weighed accurately, that is, $m = m^* = 2.981$. The other parameters are $D = 5.3062$ and $T = 0.3875$, whose actual values of D^* and T^* are difficult to know, but their ranges lie in $5.0 \leq D^* \leq 5.5$ and $0.3 \leq T^* \leq 0.4$, respectively. First, we conduct the experiments on the square responses. Figure 6.6 shows the control responses for different eddy current feedback control gains, i.e, $K_i = 0, 1, 1.2, 1.5$, with the corresponding eddy current control signal shown in Figure 6.7. The motor control signal with the gain $K = 200$ is shown in Figure 6.8. It is observed that a large overshoot exists during the first 12 s. This is due to the lack of the damping in the air bearing system. After 12 s, the eddy current is added into the system and the

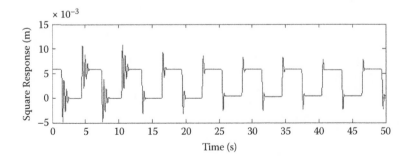

FIGURE 6.6
Square response.

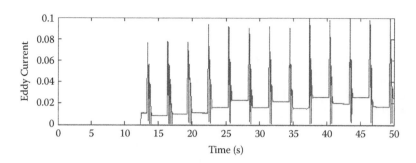

FIGURE 6.7
Eddy current control.

overshoot in response time is improved significantly. From this figure, we can find the better control gain of the eddy current feedback is 1.5.

In the second case, we consider the sine wave trajectory, which is 0.006 $\sin(wt)$m, where $w = 15$ rad/s. Thus, we obtain the bounded values of the trajectories x_{dM}, $x_{dM}^{(1)}$ and $x_{dM}^{(2)}$ as 0.006, 0.09, and 1.35, respectively. The motor control gain is $K = 200$, while the eddy current control gain is $K_i = 1.5$. Let us now check the conditions of Theorems 6.1 and 6.2. Since $K > 0$, the matrix $A = \begin{bmatrix} 0 & 1 \\ -67.0916 & -1.78 \end{bmatrix}$ is stable and $||e^{At}|| \leq e^{-0.89t}$. This implies that $\alpha = 1$, $\lambda = 0.89$. Thus, conclusion 1 of Theorem 6.1 is met, and therefore the error e_v is bounded by a given number $\epsilon = 0.05$, which is a conservative estimate. Condition 2 of Theorem 6.1 is also satisfied by computing $\lambda = 0.89 > \frac{1 \times 0.05 \times 0.3875 \times 1.5^2 \times 1}{2.981} = 0.0146$. This implies that the stability of the closed-loop system is ensured. However, in Theorem 6.1, the model uncertainty is not considered. When the parameters in the model are uncertain, we apply Theorem 6.2 to the system and check if the stability conditions are met. Since $K = 200$, $K_i = 1.5$, condition 1 of Theorem 6.2 is satisfied. This implies that the speed error e_v is bounded. By calculating $d_1 = 0.0092$, $d_2 = 0.006$, the

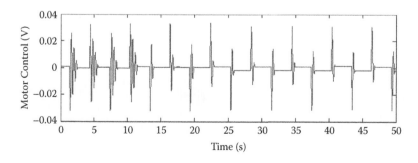

FIGURE 6.8
Motor control.

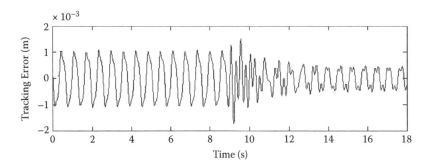

FIGURE 6.9
Tracking error.

bound of e_v is given by $|e_v| \leq B_v = max\{0.0055, 0.0265\}$. To check condition 2 of Theorem 6.2, we can compute the values of d_3 and d_4 from (6.24) and (6.26), respectively, and get $\lambda = 0.89 > \alpha \parallel b \parallel \left(\frac{T_M K_i^2}{m_m} B_v + d_2 + d_3 \right) = 0.014$. This implies that condition 2 of Theorem 6.2 is also satisfied, and therefore the error e is uniformly bounded against the model uncertainty. Furthermore, $\lim_{t \to \infty} |e| \leq \frac{0.01}{0.8760} = 0.0114$. Next, we show the experimental test. Initially, the controller is a pure feedback control without the eddy current. After 8 s the proposed controller (6.10) is applied to the system. Figure 6.9 shows the tracking performance. For two control signals, the eddy current feedback control is shown in Figure 6.10, while the motor feedback control is shown in Figure 6.11. It is observed that the tracking error under the P controller is within 1.2×10^{-3} m, while the tracking error under the proposed control (6.10) is within 5×10^{-4} m. Clearly, the tracking error is improved due to the proposed controller. This verifies our theoretical analysis. It should also be noted that the energy used for the motor control is reduced after the eddy current feedback is applied to the system.

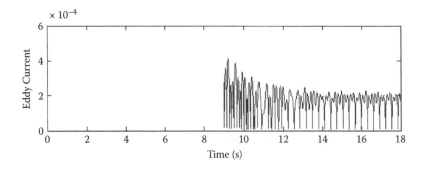

FIGURE 6.10
Eddy current control.

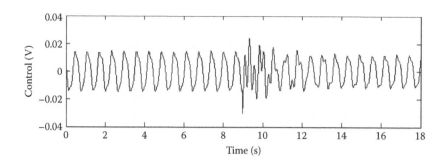

FIGURE 6.11
Motor control.

6.5 Control of Spherical Air Bearing Stage

In this section, we design a spherical air bearing stage (SABS) and focus on the control implementation for such kinds of air bearing stages. Unlike the linear air bearing stage, the SABS provides rotational motions in 2-DOF, and thus the controller design will be different from the linear air bearing stage.

6.5.1 Mechanical Structure

The mechanical structure of SABS is shown in Figure 6.12. It is cylindrical with an outer diameter of 182 mm, and mainly consists of two components: a

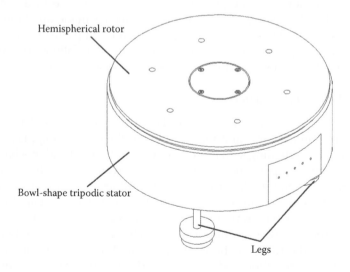

FIGURE 6.12
Mechanical structure of SABS.

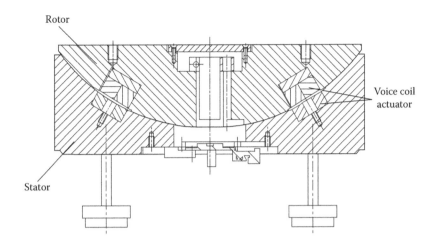

FIGURE 6.13
Cross-sectional view of SABS.

hemispherical solid rotor with a radius of 112 mm and a bowl-shaped stator for holding the rotor with three adjustable legs. The material of the stator and the rotor is aluminum alloy, which is not ferromagnetic. Figure 6.13 shows the cross-sectional view of SABS. When a thin air film is introduced between the two components with a pneumatic system, the rotor floats on the air film so that the friction between the two surfaces is extremely low, due to the low viscosity of air. Thus, the rotor can rotate in all directions freely without any resistance.

However, achieving the angular position control by only using air bearings is not possible. Therefore, a spherical motor (which is made up of four voice coil actuators) is designed, as shown in Figure 6.13. In particular, the motor is used to enable positioning in two directions. In what follows, the two key components (voice coil actuators and pneumatic system) will be elaborated.

6.5.1.1 Voice Coil Actuators

A voice coil actuator (VCA) is usually classified as a brushless DC actuator that provides linear force and motion. One of the common applications of VCA is in the positioning actuator of the read-and-write head in a hard disk drive. Due to its simple structure, it has many advantages, such as miniaturization, rapid response, high accuracy, high stiffness (i.e., high resistance to deformation), and easy to operate. Therefore, VCA is a popular choice for mechatronic equipment, especially precision positioning systems.

In the SABS, four VCAs are uniformly and symmetrically distributed inside the stator and the rotor, as shown in Figure 6.14. Each VCA has the same structure and specification, and it is composed of two parts: the coils mounted in the stator and the permanent magnets (with magnetic yokes) in the rotor. In addition, every two motors on the same axis are set as a pair; each pair is

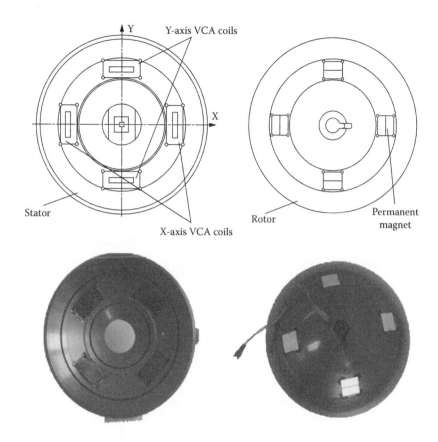

FIGURE 6.14
Stator and rotor of SABS.

used for 1-DOF angular positioning around each corresponding axis (X-axis and Y-axis), respectively. In this chapter, the pair of VCAs placed on the X-axis is named "X-axis VCA," while the other one is named "Y-axis VCA."

The working principle of each VCA is shown in Figure 6.15. More specifically, Figure 6.15(a) displays the magnetic flux path of the VCA, and Figure 6.15(b) shows the distribution of VCA. As can be seen in Figure 6.15(a), two permanent magnets made of NdFeB (neodymium iron boron) are placed side by side on one soft iron yoke, while the coils are immovable due to the fixed stator.

The permanent magnets generate a stable magnetic field. Its magnetic flux density B is a vector, which, in the Cartesian coordinate system, can be resolved into three orthogonal components: B_x, B_y, and B_z. In particular, the displacement of this VCA is quite small and the coils rarely reach the end of the magnets. Moreover, the value of B_z is minuscule, which is neglected in this case.

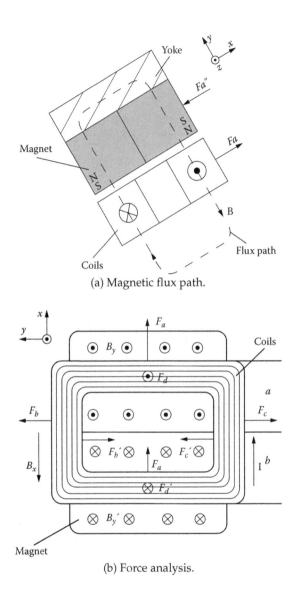

(a) Magnetic flux path.

(b) Force analysis.

FIGURE 6.15
Working principle of VCA.

When the current is applied through the coils, several forces are exerted on the coils as a result of the interaction between magnetic field and electric current, as shown in Figure 6.15(b). As can be seen, the forces F_a, F_b, F_b', F_c, and F_c' are generated in the effect B_y, where, due to the symmetrical structure, $F_b = -F_c$, $-F_b' = F_c'$, i.e., $F_b + F_b' + F_c + F_c' = 0$. Therefore, all forces along the Y-axis direction are canceled, so that the equivalent force in this direction equals zero. In other words, there is no motion generated in the Y-axis direction when the current is applied. In the X-axis direction, according to the Lorentz force law, the force F_a is given by

$$F_a = k\vec{B_y} \times \vec{L} Ni \tag{6.33}$$

where k is a constant relating to the structure of magnetic flux path, $\vec{B_y}$ is the effective flux density, $\times$ denotes vector cross-product, $\vec{L}$ is the effective length of each turn of the coils, N is the number of turns of the coils, and i is the applied current. And Equation (6.1) can be rewritten as below:

$$F_a = kB_yLNi \tag{6.34}$$

where B_y and L are scalars.

Meanwhile, arising from the force F_a, an equal but opposite force F_a'' is produced on the magnets. Because the coils are fixed as above mentioned, the magnet part will be driven under the action of the force F_a'' in the X-axis direction when the coils are subjected to current.

Besides the effect of B_x, a couple of forces F_d and F_d' in the Z-axis direction are generated, which are approximately equal in magnitude and opposite in direction but not collinear. Thus, the equivalent force along the Z-axis also equals zero (because $F_d = -F_d'$), but a turning torque M_d around the Y-axis is inevitably generated:

$$M_d \approx F_d d \tag{6.35}$$

where d is the distance between F_d and F_d'.

Actually, the torque M_d is too small to affect the performance and stability of SABS since the value of d is relatively small. Therefore, this torque can be ignored in this system.

In conclusion, the VCA mainly generates a force in the tangential direction of rotor when the coils are energized. Thus, by controlling the current of different pair of VCAs, the spherical motor is able to implement angular positioning in two-DOF, where, the X-axis VCA is used for adjusting the angle around Y-axis (yaw), while the Y-axis VCA is for the angle around X-axis (roll).

6.5.1.2 Pneumatic System

The pneumatic system of SABS includes the spherical air bearing and other components that are used to provide pressurized air. For air bearing, in terms of the pressure generation principles, it can be categorized as: (1) hydrodynamic (or self-acting) type—air film is generated internally by relative motion at high speeds; (2) hydrostatic type—air film is provided from external pressure supply; and (3) squeeze-film type—air film is created by imposing oscillations on the nonmoving part of bearings. Generally, the hydrodynamic type is common for fluid bearings in which viscosity of fluid (such as oil) is relatively high, so it is hard for air bearings to utilize this type due to the low viscosity of air. The squeeze-film type is frequently used for damping applications. Besides, it is easy to obtain and control a designated air pressure by an external air compressor. As a result, the hydrostatic type is applied in the SABS.

In order to maximize the stiffness of the air bearing and help it maintain a constant air gap, it is better to preload the air bearing, although air bearing can work without any preload. In the SABS, a vacuum preload is mainly used, which avoids unnecessary moving mass for preloading in the system.

The structure and the working principle schematic diagram of hydrostatic air bearing in the SABS are shown in Figure 6.16. As can be seen, eight microholes (not all shown in the figure) distributed in two circles uniformly and symmetrically are used to transmit compressed air into the floating area, and in turn generating forces for sustaining the rotor. One intake is used to evacuate gas so as to form the vacuum room, which provides vacuum preloading. The forces generated by the air pressure can be calculated by Equation (6.36):

$$F_P = PA \qquad\qquad (6.36)$$

where P is the average pressure in the floating area or vacuum room, and A is the equivalent area of those areas.

Since SABS is a hydrostatic type, external clean pressurized air is required for the bearing (i.e., air compressor and filters are required). Additionally, due to the vacuum preload method, the air bearing needs a vacuum generator. Those components and the air bearing compose the whole pneumatic system, as shown in Figure 6.17.

6.5.2 Control System

Controlling an air bearing stage can be defined as influencing it in such a way as to force it to operate according to certain requirements. Moreover, the control objective of the SABS is to find a control mechanism for every bounded smooth desired output so that the controlled output converges to the desired output as closely as possible. This involves system modeling, parameter identification, eliminating noise, and controller design.

6.5.2.1 Hardware of Control System

A control system for the SABS is developed to control each VCA in order to implement angular positioning control. The control system and mechanical components together compose the SABS. The system diagram is shown in Figure 6.18, which includes a sensor, motor drivers, and a controller.

The sensor is a dual-axis position sensing diode (PSD), the outputs of which are bipolar voltage analogs of the X and Y positions of the light spot centroid. The light spot is provided by a compact laser generator that is mounted on

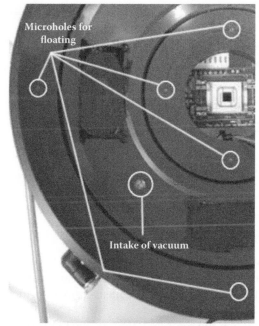

(a) Arrangement of air vents.

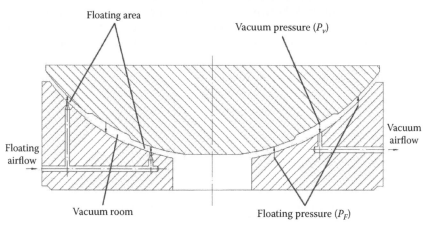

(b) Design scheme and working principle.

FIGURE 6.16
Structure and working principle of air bearing in the SABS.

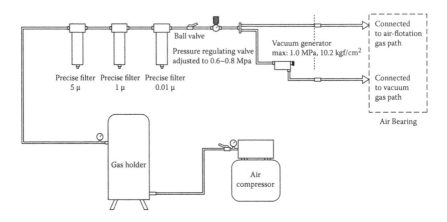

FIGURE 6.17
Pneumatic system of SABS.

the rotor. Moreover, the PSD exports the measured values of the X and Y positions separately. Therefore, the angles of roll and yaw can be calculated by the sensor outputs, respectively. The measurement principle for angles is shown in Figure 6.19.

As can be seen, when the rotor is stationary, the light spot and the center of PSD are coincident, and therefore the outputs are zero. When the rotor rotates around its pivot, an angle between the axis of rotor (which can be represented by the laser beam) and the direction of gravity is formed. Meanwhile, the distances between the light spot and the PSD central point are measured by the PSD. Thus, according to the trigonometric function, the angles can be calculated by Equation (6.37) as follows:

$$\left. \begin{aligned} \theta &= \arctan\left(\frac{y}{h}\right) \\ \varphi &= \arctan\left(\frac{x}{h}\right) \end{aligned} \right\} \tag{6.37}$$

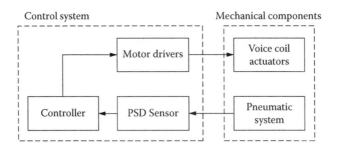

FIGURE 6.18
System diagram of SABS.

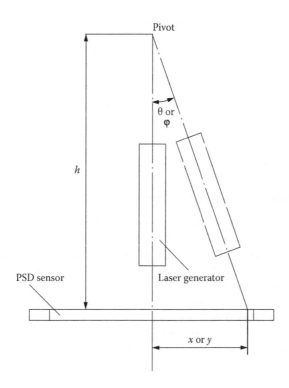

FIGURE 6.19
Angles measurement principle.

where θ and φ are the angles of roll and yaw, x and y are the outputs of the PSD, and h is the height between the PSD and the rotor's pivot.

Since the angle outputs of the SABS are designed to be less than $\pm 10°$, Equation (6.37) can be linearized and written as

$$\left. \begin{array}{c} \theta \approx \dfrac{y}{h} \\[2mm] \varphi \approx \dfrac{x}{h} \end{array} \right\} \tag{6.38}$$

Two motor drivers are required to drive two pairs of VCAs, respectively. Each drive consists of one high-voltage, high-current operational amplifier (OPA 548), which controls the current output in terms of the analog voltage control signal. The continuous output current of OPA 548 is 3 A, while the peak output current is 5 A, which meets the requirements of the VCAs.

The controller is the center part of the control system. In the SABS, a dSPACE 1104 control development card and a personal computer (PC) are used to implement the positioning control.

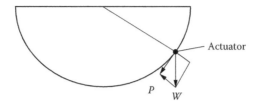

FIGURE 6.20
Force caused by actuator.

6.5.2.2 Modeling of Air Bearing Stage

The use of the air bearing stage for an accurate control depends greatly on the model obtained for the controller. The purpose of the model presented in this subsection is to map the relationship between input and output at the electrical port of the air bearing system in a special form that can be represented by a set of equations.

In the spherical system, there are two pairs of voice coil actuators that can control X–Y directions. These actuators are placed inside the stator and rotor as mentioned before (see Figure 6.14). According to the theory of VCA, the force generated by the actuator is given by

$$F = kB_yLNi$$

Applying the physical law, the SABS can be described by the following mathematical equations:

$$\left.\begin{array}{l} m_1\ddot{x}_1 = F_1 - P_1 \\ m_2\ddot{x}_2 = F_2 - P_2 \end{array}\right\} \tag{6.39}$$

where $P_r, r = 1, 2$, is a force caused by the weight of the air bearing stage (see Figure 6.20), $x_1 = \theta R$, and $x_2 = \varphi R$, where R is the rotor's radius of sphere. We assume that the actuators on X-Y are decoupled.

6.5.2.3 Parameter Identification

Though the model structure is built in the above subsection, the model parameters P_r and $m_r, r = 1, 2$, are difficult to obtain. In this subsection, we design an estimator to obtain the parameters of the system. The idea of the algorithm is based on adaptive control concepts.

Consider the control problem of the system in Equation (6.39). The definition of the filtered error S is given by

$$S = \lambda e + \dot{e} \tag{6.40}$$

where λ is a constant and the error $e = x_d - x$, x_d represents the desired performance and $x = [x_1\ x_2]^T$ represents the state variables of the system.

Thus, the system can be rewritten as

$$\dot{S} = \lambda\dot{e} + \ddot{x}_d + \frac{P}{m} - \frac{kB_yLNi}{m} = \lambda\dot{e} + \ddot{x}_d + a - bi \qquad (6.41)$$

where $a = \frac{P}{m}$ and $b = \frac{kB_yLN}{m}$.

Since the parameters a and b are not known exactly, the following adaptive control law is used

$$i = \frac{K_vS + \lambda\dot{e} + \ddot{x}_d + \hat{a}}{\hat{b}} \qquad (6.42)$$

with

$$\dot{\hat{a}} = \gamma_1 S \qquad (6.43)$$

$$\dot{\hat{b}} = -\gamma_2 i S \qquad (6.44)$$

Substituting the control law into the system yields

$$\dot{S} = \tilde{a} - \tilde{b}i - K_vS \qquad (6.45)$$

Define $V = S^2 + \frac{1}{\gamma_1}\tilde{a} + \frac{1}{\gamma_2}\tilde{b}$. The time derivative of V is given by

$$\dot{V} = -2K_vS^2 + 2S\tilde{a} - 2S\tilde{b}i - \frac{2}{\gamma_1}\tilde{a}\dot{\hat{a}} - \frac{2}{\gamma_2}\tilde{b}\dot{\hat{b}} \qquad (6.46)$$

Substituting the adaptive learning laws produces

$$\dot{V} = -2K_vS^2 \qquad (6.47)$$

This shows S and $\hat{b}$, $\hat{a}$ are bounded. This also implies that $\dot{S}$ is bounded. From (6.47), we have

$$\int_0^t K_vS^2 d\tau \le V(0) - V(t) \qquad (6.48)$$

Applying Barbalat's lemma, we obtain

$$\lim_{t\to\infty} ||S||_2 = 0 \qquad (6.49)$$

Although the online parameter estimation scheme guarantees that the tracking error S converges to zero as $t \to \infty$, this result is not sufficient to establish parameter convergence to the true parameter values unless the input signal i is sufficiently rich. The system noise (white noise) contributes toward the cause.

6.5.2.4 Noise Filter

In the SABS, noise is introduced due to the sensor, circuit, and amplifier. To analyze the signal noise, we adopt a fast Fourier transform (FFT) algorithm.

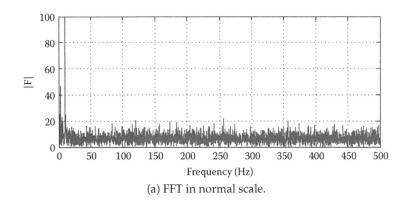

(a) FFT in normal scale.

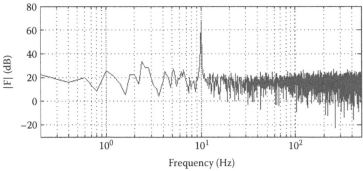

(b) FFT in logarithmic scale.

FIGURE 6.21
Spectrum of the SABS.

It is a popular offline approach widely used to obtain information of the frequency distribution required for the filter design. Figure 6.21 shows the spectrum of the air bearing stage, where the frequency of 10 Hz is a working signal. It is observed that the noise exists over a wide range with small amplitude. Moreover, the expected value of the noise approaches zero. Since it exhibits the characteristics of white noise, an observer-type filter is used for handling such noise.

In the presented state space model, the position of the air bearing stage is available. Rewriting (6.39) into a compact form yields

$$\left.\begin{aligned} \dot{X} &= AX + B(i - a) + w \\ y &= \qquad CX + v \end{aligned}\right\} \tag{6.50}$$

where $X = [x \; \dot{x}]^T$, y is the output (position), w and v are the system noises, and

$$A = \begin{bmatrix} 0 & 1 \\ 0 & 0 \end{bmatrix}, B = \begin{bmatrix} 0 \\ b \end{bmatrix}, C = [1 \; 0] \tag{6.51}$$

$$a = \frac{P}{m}, b = \frac{kB_yLN}{m} \tag{6.52}$$

For the noise, the following conditions are assumed:

$$(i) \quad E(w) = 0, E(ww^T) = Q_c\delta, \tag{6.53}$$

$$(ii) \quad E(v) = 0, E(vv^T) = R_c\delta \tag{6.54}$$

$$(iii) \quad E(wv^T) = 0 \tag{6.55}$$

where E(.) denotes the expectation and δ is the Dirac delta function.

Because both output and state have noise contents, the observed outputs are used to replace the actual outputs. The following observer-type filter is designed:

$$\left.\begin{aligned} \dot{\hat{X}} &= A\hat{X} + B(i - a) + K(y - \hat{y}) \\ \hat{y} &= \qquad\qquad C\hat{X} \end{aligned}\right\} \tag{6.56}$$

In (6.56), K is the steady-state gain of the filter. The estimated error is given by

$$\dot{\tilde{X}} = \bar{A}\tilde{X} + w - Kv \tag{6.57}$$

where $\tilde{X} = X - \hat{X}$, $\bar{A} = A - KC$. Basically, the gain vector K is required to be designed to satisfy the condition: all the eigenvalues of $\bar{A}$ have negative real parts, i.e., $Re(\lambda\{\bar{A}\}) < 0$. Moreover, the criterion for determining K is the error between the observer and actual outputs as small as possible.

6.5.2.5 Model-Based Controller Design

In this subsection, we will design a proportional-integral-derivative (PID) controller. The main reason is that it has a simple structure, which is easily understood by engineers, and works well under practical conditions.

The model-based PID controller uses the following law:

$$i = K_p(y_d - \hat{y}) + K_I \int_0^t (y_d - \hat{y})d\tau + K_D\frac{d(y_d - \hat{y})}{dt} \tag{6.58}$$

It should be noted that $\hat{y}$ is the observed output. Figure 6.22 shows the schematic diagram of the overall model of a PID controller. Unlike traditional PID control, this model-based PID control is based on the observer output

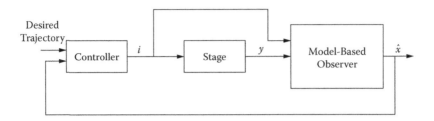

FIGURE 6.22
The overall structure of PID control.

that is designed based on the system model. The PID parameters are tuned based on a trial-and-error method.

6.6 Performance Analysis of Spherical Air Bearing System

Evaluation of the state of the SABS, in terms of whether or not it fulfills the assumed requirements, is done on the basis of the system control performance. Moreover, it is done on the basis of observations of the position variable and characterizing the process behavior. This is a real-time experimental study. The block diagram of the whole closed-loop control system is shown in Figure 6.23. The setup of SABS is shown in Figure 6.24. This involves air bearing, sensor, drivers, and controller. The controller is designed based on a dSPACE control development card, which utilizes the Texas Instruments TMS320C31 32-bit floating point processor. The control algorithm is implemented via MATLAB/Simulink block diagrams with RTI (Real-Time Interface), compiled on a PC that can be downloaded into the dSPACE board.

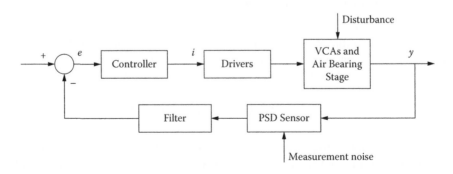

FIGURE 6.23
Block diagram of control system.

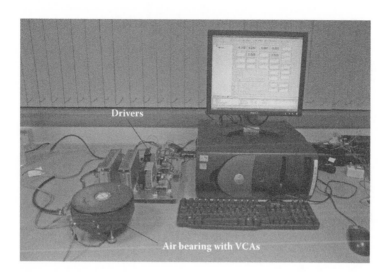

FIGURE 6.24
Setup of SABS.

6.6.1 Model Identification

We first identify the model parameter of the air bearing motor from the entire system. The range of the desired signal used is chosen as 1 Hz. The proportional-derivative (PD) control gain is chosen as $K_P = 0.1$, $K_D = 0.001$. In this experiment, the initial values of the parameters in the adaptive laws (6.43) and (6.44) are chosen as $[\hat{b}(0), \hat{a}(0)] = [0.08, 0.0001]$. The adaptation rates in the laws are selected as $\gamma_1 = 0.1$, $\gamma_2 = 0.1$. The time evolution of the parameter estimates using the proposed adaptive algorithm is shown in Figure 6.25. The parameters $\hat{b}$ and $\hat{a}$ converge to 0.0755 and 0.0001, respectively. It is observed that during the initial learning phase, the estimated parameter $\hat{b}$ converges the fastest. This is due to the lack of knowledge about the plant. Through adaptive learning, the parameters converge to their true values after 230 s. Finally in this experiment, the values of a and b are identified approximately, but they are approaching the actual values, since the system noise is almost white noise, which adds to the excitation of signal partially.

6.6.2 Noise Filter

Based on the model (6.50), the filter (6.56) is designed, that is,

$$\dot{\hat{X}} = \begin{bmatrix} 0 & 1 \\ 0 & 0 \end{bmatrix} \hat{X} + \begin{bmatrix} 0 \\ 0.0755 \end{bmatrix} (i - 0.0001) + K(y - \hat{y})$$

$$\hat{y} = [1\ 0]\hat{X}$$

(6.59)

The choice of the gain K depends on the condition. Initially, we choose $K = \begin{bmatrix} 2 \\ 4 \end{bmatrix}$. Then, we change to $K = \begin{bmatrix} 5 \\ 25 \end{bmatrix}$ to compare the filter performance.

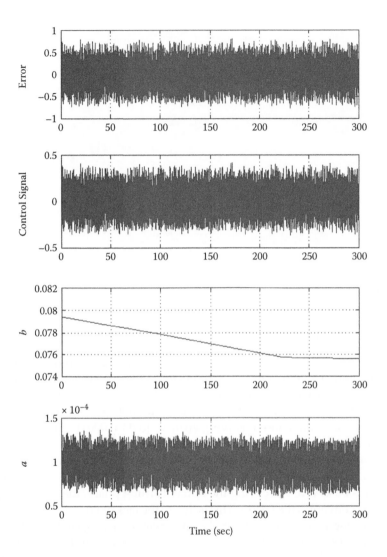

FIGURE 6.25
Model identification.

Figure 6.26 shows the filter performance compared with the actual sine wave signal. It is found that for the case of $K = \begin{bmatrix} 2 \\ 4 \end{bmatrix}$ ($0 \sim 22$ s), the estimated value is not satisfactory, while for the case of $K = \begin{bmatrix} 5 \\ 25 \end{bmatrix}$ (after 22 s), the estimated value converges to the actual sine wave signal. To show the convergence clearly, we plot the figure under a small time window ($30 \sim 40$ s), and the result is shown in Figure 6.27. It is seen from the figure that the noise is removed by the

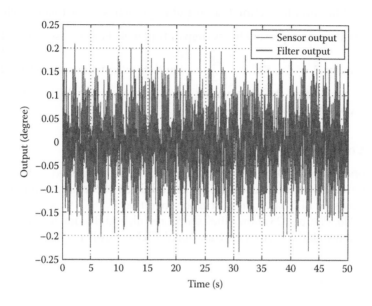

FIGURE 6.26
Filter performance analysis (I).

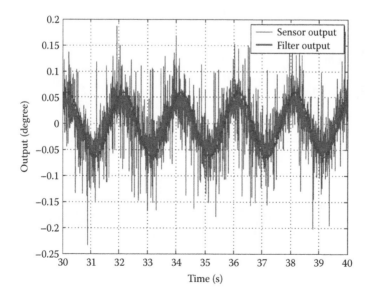

FIGURE 6.27
Filter performance analysis (II).

designed filter and the estimated output approaches the actual one closely. Now, this filter can represent the system output well. For the case of $K = \begin{bmatrix} 5 \\ 25 \end{bmatrix}$, we can obtain $\bar{A} = A - KC = \begin{bmatrix} -5 & 1 \\ -25 & 0 \end{bmatrix}$ and its eigenvalues are $-2.5 \pm 4.33i$. The real parts of all the eigenvalues of $\bar{A}$ are negative, which satisfies the stability condition: $Re(\lambda\{(\bar{A})\}) < 0$.

6.6.3 Control Results

The model-based PID controller in Equation (6.58) is applied to the SABS. The parameters of the PID controller are adjusted according to the general tuning rules for PID parameters. It is obtained that $K_p = 0.01$, $K_I = 0.12$, and $K_D = 0.008$. Figure 6.28 shows the control result, and the maximum position error is about $0.0125°$. If using traditional PID controller without observer,

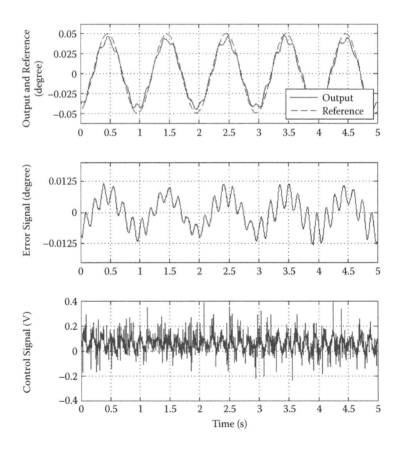

FIGURE 6.28
Model-based PID control.

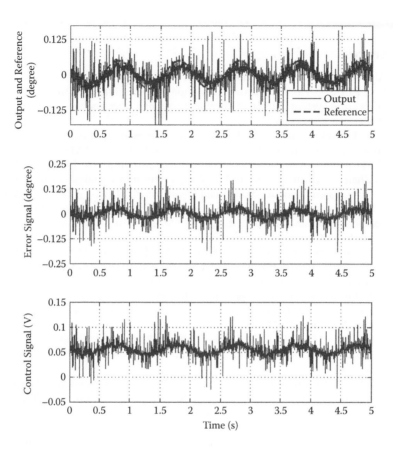

FIGURE 6.29
Traditional PID control.

the maximum error is about 0.25° and the result is shown in Figure 6.29. It is clearly observed from the figures that the tracking error is significantly improved by the proposed control.

6.7 Concluding Remarks

This chapter has presented the control algorithms for two types of air bearing systems. The first one is linear air bearing stage. The control design is based on the P control with eddy current braking. The stability of the closed-loop system has been proven rigorously. The second type is a spherical air bearing stage, wherein a filtered control approach is proposed to overcome the system noise and achieve optimal performance. Experimental results have verified the effectiveness of the proposed control schemes.

References

1. Air bearing principle. Brochure, Hovair Systems Inc.
2. K. C. Fan, C. C. Ho, and J. I. Mou. Development of a multiple microhole aerostatic air bearing system. *Journal of Micromechanics and Microengineering*, 12, 636–543, 2002.
3. D. E. Ezenekwe, and K. M. Lee. Design of air bearing system for fine motion application of multi-DOF spherical actuators. In *1999 IEEE/ASME International Conference on Advanced Intelligent Mechatronics*, Atlanta, GA, 1999, pp. 812–818.
4. Q. Zhang, X. Shan, G. Guo, and S. Wong. Performance analysis of air bearing in a micro system. *Materials Science and Engineering A*, 423(1–2), 225–229, 2006.
5. J. L. Schwartz and C. D. Hall. System identification of a spherical air-bearing spacecraft simulator. Presented at *2004 AAS/AIAA Space Flight Mechanics Meeting*, Maui, HI, 2004.
6. A. Sescu, C. Sescu, F. Dimofte, S. Cioc, A. A. Afjeh, and R. Handschuh. A study of the steady-state performance of a pressurized air wave bearing at concentric position. Tribology Transactions, 52(4), 544–552, 2009.
7. K. K. Tan and S. Huang. Problem and solution of designing an air bearing system. In *2010 2nd International Conference on Industrial and Information Systems*, Dalian, China, 2010, pp. 212–215.
8. S. Q. Lee and D. G. Gweon. A new 3-DOF Z-tilts micropositioning system using electromagnetic actuators and air bearings. *Precision Engineering*, 24(1), 24–31, 2000.
9. D. J. Lee, K. Kim, K. N. Lee, H. G. Choi, N. C. Park, Y. P. Park, and M. G. Lee. Robust design of a novel three-axis fine stage for precision positioning in lithography. *Proceedings of the IMechE, Part C: Journal of Mechanical Engineering Science*, 224(4), 877–888, 2010.
10. K. Lee and K. Park. Optimal robust control of a contactless brake system using an eddy current. *Mechatronics*, 9(6), 615–631, 1999.
11. X. Xie, Z. Wang, P. Gu, Z. Jian, X. Chen, and Z. Xie. Investigation of magnetic damping on an air track. *American Journal of Physics*, 74(11), 974–978, 2006.
12. K. K. Tan, S. Huang, C. S. Teo, and R. Yang. Damping estimation and control of a contactless brake system using an eddy current. Presented at Proceedings of the 8th IEEE International Conference on Control and Automation (ICCA2010), Xiamen, China, June 9–11, 2010.

7

Fault Detection and Accommodation in Actuators

The precision requirements associated with precise actuators are high with low tolerance for fault and change in characteristics over time. To ensure safe and efficient operation of control systems against various failures and prolong achievable performance, a failure detection and accommodation scheme is essential. If such a detection cannot be put in place, total shutdown of the manufacture may occur, resulting in lost revenues to the operating utility. The development of an efficient diagnostic strategy is thus one of the most important and challenging problems in control engineering. Over the past two decades considerable research efforts have been done to detect changes whenever a failure occurs. The problem of fault diagnosis is to detect failures in a physical system by monitoring its input and output information. The main goal of a fault diagnosis system is to monitor the system during its normal working conditions so as to detect the occurrence of failures and recognize the fault types.

Various approaches to fault detection have been reported over the last two decades. In [1], a parameter estimation is used to diagnose faults in a DC motor. In [2], a chip thickness and cutting force model is built for predicting process faults. In [3], a technique to improve the fault detection is presented by using the classical multiple signal classification (MUSIC) method. In [4,5], we develop linear state observers for detecting the cutting tool wear and monitoring the motor system. However, the model-based fault detection schemes depend on the assumption that a mathematical characterization of the system is available. In practice, this may not hold since it is difficult to obtain an accurate model. Recently, nonlinear approximation approaches to the fault detection problem have been developed [6–10]. Furthermore, in many applications, it is important not only to diagnose but also to accommodate a failure as quickly as possible. The works of fault-tolerant control (FTC) can be found in [11,12]. Most of these studies are focused on single-input-single-output (SISO) systems. The problem that arises in multiple-input-multiple-output (MIMO) systems introduces additional complexities. The work of [13–15] is focused on control of MIMO linear systems with actuator or state failures.

In this chapter, fault diagnosis and accommodation control schemes for a general MIMO precise actuator system are presented. There are three main features: (1) the stability analysis of the proposed schemes are investigated for two different operating modes—in the absence of faults and after fault detection; (2) both the state feedback and output feedback control schemes are discussed; and (3) the real-time testing is conducted based on a real machine actuation system.

7.1 Problem Statements

Consider the following MIMO precise actuator system described by

$$\ddot{q} = M^{-1}(q)[u - V_m(q,\dot{q}) - H(\dot{q}) - H_G(q) - \tau_d] + \mathcal{B}(t - T)\varsigma(x) \quad (7.1)$$

where $q = [q_1, q_2, \ldots q_m]^T \in R^m$ are the joint positions of the subsystems, $M(q)$ are the symmetric positive definite inertia matrices, $V_m(q,\dot{q})$ represent Coriolis and centripetal forces, $H(\dot{q})$ are the dynamic frictional force matrices, τ_d are load disturbance matrices, $H_G(q)$ are the potential energy terms, and u denote generalized input controls of the system applied at the joints.

Consider faults with time profiles modeled by

$$\mathcal{B}(t - T) = diag\{\beta_1(t - T), \beta_2(t - T), \ldots, \beta_m(t - T)\}$$

$$\beta_i(t - T) = \begin{cases} 0 & t < T \\ 1 - e^{-\theta_i(t-T)} & t \geq T \end{cases} \quad (7.2)$$

where the fault occurrence time T is unknown, and $\theta_i > 0$ is an unknown constant that represents the rate at which the fault in states and actuators evolves. Since the friction matrix $H(\dot{q})$ may be unknown accurately, a nominal matrix is defined as $\bar{H}(\dot{q})$, and then the uncertain matrix is $H(\dot{x}) - \bar{H}(\dot{q})$.

The system (7.1) can also be written as a compact form:

$$\left. \begin{aligned} \dot{x} &= A_0 x + b[F(x) + G(q)u + \eta(x, t) + \mathcal{B}(t - T)\varsigma(x)] \\ y &= Cx \end{aligned} \right\} \quad (7.3)$$

where

$$x = [q_1, \dot{q}_1, q_2, \dot{q}_2, \ldots, q_m, \dot{q}_m]^T \quad (7.4)$$

$$A_0 = diag\{A_{01}, A_{02}, \ldots, A_{0m}\} \quad (7.5)$$

$$b = diag\{b_1, b_2, \ldots, b_m\} \quad (7.6)$$

$$F = -M^{-1}(q)[V_m(q,\dot{q}) + \bar{H}(\dot{q}) + H_G(q)] \quad (7.7)$$

$$G = M^{-1}(q), \eta = -M^{-1}(q)[H(\dot{q}) - \bar{H}(\dot{q}) + \tau_d] \quad (7.8)$$

$$C = diag\{C_1, C_2, \ldots, C_m\} \quad (7.9)$$

$$A_{0i} = \begin{bmatrix} 0 & 1 \\ 0 & 0 \end{bmatrix}, b_i = \begin{bmatrix} 0 \\ 1 \end{bmatrix}, C_i = [1, 0] \quad (7.10)$$

This chapter has the following two objectives: (1) detect a fault when the monitored system fails to function normally, and (2) after a fault is detected, require that the controller should be reconfigured to accommodate the fault. The basic assumptions for the problems stated are

1. The modeling uncertainty $\eta(\mathbf{x}, t)$ is bounded by a known constant, i.e.,

$$||\eta(\mathbf{x}, t)|| \leq \bar{\eta} \tag{7.11}$$

2. The desired trajectories $y_d = [y_{d1}, y_{d2}, \ldots, y_{dm}]^T$ are known bounded functions of time with bounded known derivatives.

7.2 Types of Failure

Understanding of the machine behavior in healthy state and under fault conditions is the essence of any reliable fault diagnosis technique. Since an actuator system relies on actuator motion and measured electrical sensors, there are two main classes of failures: actuator and sensor failures. The former may result in severe damage to the system, while the latter leads to a mistaken operation. Several examples are given below to illustrate possible failures in an actuator system.

7.2.1 Examples of Actuator Failure

Mechanical failure: Consider an H-type stage with two parallel linear motors (see Figure 7.1). A mechanical failure arises when in a moving

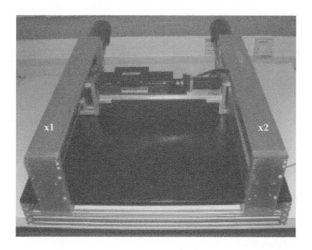

FIGURE 7.1
A mechanical failure arising due to unsynchronized motion.

H-type stage the two linear motors (x_1 and x_2 of Figure 7.1) are not synchronized in motion.

Short circuits: Short circuits can occur when the insulation of the wiring used breaks down. Short circuits can even occur when electric motors are forced to operate when the moving parts are jammed. This can result in abnormal buildup of current, ultimately leading to a short circuit [17].

Coil failure: Many factors can result in such a failure. The two main factors are the skin and proximity effects, which lead to a nonuniform current distribution within a copper coil. Areas of high current density within this distribution are the primary candidates for localized hot spots, which can result in premature coil failure [18].

7.2.2 Sensor Failure

Encoder failure: In a linear actuator system, the encoder sensor is used to determine the position of an object in a linear actuator. This information can be used to control the linear actuator. However, the encoder is vulnerable to one type of damage: seal failure, which permits the entry of contaminants (oil, dirt, water, etc.). Why do contaminants cause encoder failure? See Figure 7.2. An encoder sensor consists of a code disk and infrared light. The sensor shines a light through the disk and detects variations in the amount of light. When there is any type of contamination on the disk, it breaks the light and thus creates errors, for the sensor needs to see tiny lines on the disk accurately.

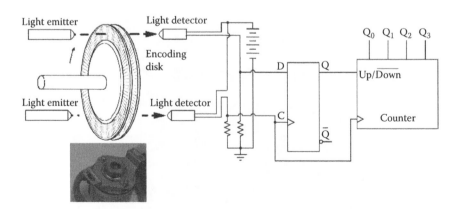

FIGURE 7.2
Working principle of encoder.

7.3 Fault Diagnosis Scheme

7.3.1 Fault Detection

For the fault detection algorithm, we consider the following nonlinear model as an estimator:

$$\dot{\hat{x}} = A_0\hat{x} + b[F(x) + G(q)u] + L_{ox}\tilde{x} \tag{7.12}$$

$$\hat{y} = C\hat{x} \tag{7.13}$$

where $\hat{x}$ denotes the estimated state vector x, the variable $\tilde{x}$ is defined as $x - \hat{x}$, and L_{ox} is a constant matrix. The next step in the construction of the fault detection scheme is the design of the algorithm for monitoring a fault occurrence.

Combining (7.12) with (7.3), the following error equation is produced:

$$\dot{\tilde{x}} = \bar{A}_0\tilde{x} + b[\eta(x, t) + B(t - T)\varsigma(x)] \tag{7.14}$$

where $\bar{A}_0 = A - L_{ox}$. Note that L_{ox} should be designed so that $\bar{A}_0$ is stable. Based on the error $\tilde{y}_i(t) = y_i - \hat{y}_i$, $i = 1, 2, \ldots, m$, a fault estimation algorithm is presented. Since $B(t - T)\varsigma(x)$ is zero when $t < T$, the solution of the state estimation error between (7.3) and (7.12) is given by

$$\tilde{y}_i(t) = \bar{C}_i e^{\bar{A}_0 t}\tilde{x}(0) + \bar{C}_i \int_0^t e^{\bar{A}_0(t-\tau)}b\eta(x, \tau)d\tau, \quad t < T \tag{7.15}$$

where

$$\bar{C}_i = diag\{0, \ldots, C_i, \ldots, 0\} \tag{7.16}$$

The time-varying threshold bound ϖ_i is chosen as follows:

$$\varpi_i = ||\bar{C}_i e^{\bar{A}_0 t}\tilde{x}(0)|| + \int_0^t ||\bar{C}_i e^{\bar{A}_0(t-\tau)}b||\bar{\eta}(x, \tau)d\tau \tag{7.17}$$

The decision that a fault has occurred is made when at least one component of the estimation error $|\tilde{y}_i(t)|$ exceeds its corresponding threshold bound ϖ_i. The fault detection time is defined as

$$T_d = min\{t_i \geq T \mid |\tilde{y}_i(t_i)| > \varpi_i, i = 1, 2, \ldots m\} \tag{7.18}$$

7.3.2 Fault Isolation

For a practical problem, even if the occurrence of a fault has been detected, it is necessary to isolate the faulty element and find out the fault types. As the fault is unknown, the fault isolation task may require all possible fault functions for finding a fault type (or fault pattern). Define a fault function set Ω_f that includes all possible faults,

$$\Omega_f = \{\theta_0^f \zeta_0^f(t), \theta_1^f \zeta_1^f(t), \ldots, \theta_s^f \zeta_s^f(t)\} \tag{7.19}$$

where $\theta_r^f = [\theta_{r,ij}^f], r \in [0, s], i \in [1, m], j \in [1, n]$ within $\theta_{r,ij}^{fm} \le \theta_{r,ij}^{f} \le \theta_{r,ij}^{fM}$. The idea of the isolation scheme is to use the multiple observer method as suggested in [16]. Thus, the following $s+1$ isolation estimators corresponding to one of the possible types of faults are given by

$$\dot{\hat{x}}_r = A_0 \hat{x}_r + b[F(x) + G(q)u + \bar{\theta}_r \zeta_r^f(t)] + L_{ox} \tilde{x}_r, \quad r \in [0, s] \tag{7.20}$$

where $\hat{x}_r$ is the estimate of x, $\bar{\theta}_r \zeta_r^f(t)$ is the estimated fault, and $\tilde{x}_r = x - \hat{x}_r$. To derive threshold values for the proposed isolation algorithm, we consider the case in the presence of the rth fault, i.e., $\mathcal{B}(t - T)\zeta = \theta_r^f \zeta_r^f(t)$. With (7.3) and (7.20), the error dynamical equation is given by

$$\dot{\tilde{x}}_r = \bar{A}_0 \tilde{x}_r + b[(\theta_r^f - \bar{\theta}_r)\zeta_r^f(t) + \eta(x, \tau)] \tag{7.21}$$

In this case, each residual signal is given by

$$\tilde{y}_{ir}(t) = \bar{C}_i e^{\bar{A}_0 t} \tilde{x}_r(0) + \bar{C}_i \int_0^t e^{\bar{A}_0(t-\tau)} b[(\theta_r^f - \bar{\theta}_r)\zeta_r^f + \eta] d\tau$$

The threshold of this signal is given by

$$\varpi_{ir} = ||\bar{C}_i e^{\bar{A}_0 t} \tilde{x}_r(0)|| + \int_0^t ||\bar{C}_i e^{\bar{A}_0(t-\tau)} b||(\mathcal{O}_M ||\zeta_r^f|| + \bar{\eta}) d\tau$$

where

$$\mathcal{O}_M = \sqrt{\sum_i \sum_j \left[\frac{\theta_{r,ij}^{fM} - \theta_{r,ij}^{fm}}{2} + |\bar{\theta}_{r,ij}| - \frac{\theta_{r,ij}^{fm} + \theta_{r,ij}^{fM}}{2} \right]^2}$$

The isolation is made when the estimation error $|\tilde{y}_{ir}(t)|$ is less than its corresponding threshold bound ϖ_{ir}.

7.3.3 Fault Identification

When a fault cannot be isolated, fault identification has to be used to find the fault characteristics. Since the fault function is unknown, an estimator must be designed to identify the fault.

The following estimated model of the form is considered:

$$\dot{\hat{x}}_I = A_0 \hat{x}_I + b[F(x) + G(q)u] + L_{ox} \tilde{x}_I + b\hat{\zeta}(x) \tag{7.22}$$

$$\hat{y}_I = C\hat{x}_I \tag{7.23}$$

where $\hat{x}_I$ denotes the estimated state vector x_I, $\tilde{x}_I = x - \hat{x}_I$, and L_{ox} is a constant matrix. Here, $\hat{\zeta}$ is a nonlinear function that can be determined by

$$\hat{\zeta}(x) = \hat{\Xi}^T \Phi(x) \tag{7.24}$$

where $\Phi(x)$ may be fuzzy logic, polynomial series, or neural network basis.

The following adaptive law for the weights of the function basis is applied to the system:

$$\dot{\hat{\Xi}} = \Upsilon \Phi(x) D[e] - \eta ||D[e]|| \hat{\Xi} \tag{7.25}$$

where $\Upsilon = \Upsilon^T$ is a positive definite adaptation matrix and $D[.]$ is the dead-zone operator, defined as

$$D[e] = \begin{cases} 0_l & \text{if} ||e|| \leq \varpi \\ e & \text{otherwise} \end{cases} \tag{7.26}$$

where 0_l is a one-dimensional vector of zeros. The initial weight vector $\hat{\Xi}(0)$ is chosen such that $\hat{\Xi}^T \Phi = 0$ corresponding to the case without failures. This can be achieved by simply setting the weights of the function output to zero.

The fault function identified can be used for finding the failure mode by comparing it with any known failure mode. If such failure function cannot be found, this failure can be stored in a postfailure model base. Some inference rules and associative memories can be provided to match the failure mode.

7.4 Control of the System under No-Fault Condition

Before the fault occurrence, a controller for the system (7.1) is considered when there is no fault.

For given desired trajectories $y_{di}(t) \in R$, the position errors are defined as $e_i(t) = y_i - y_{di}$. Then, the filtered tracking errors are given by

$$s_1 = \left(\frac{d}{dt} + k_1\right)e_1, \ldots, s_m = \left(\frac{d}{dt} + k_m\right)e_m \tag{7.27}$$

where $k_1, \ldots, k_m$ are positive constants to be selected. Thus, the system equation can be written as

$$\dot{S}(t) = F(x) + G(q)u + v - y_d^{(2)} + n(x, t) + B(t - T)\zeta(x) \tag{7.28}$$

where

$$S = [s_1, s_2, \ldots, s_m]^T$$
$$y_d^{(2)} = [y_{d1}^{(2)}, y_{d2}^{(2)}, \ldots y_{dn}^{(2)}]^T$$
$$v = [v_1, v_2, \ldots, v_m]^T$$

with $v_i = k_i \dot{e}_i$.

In the absence of faults, the original system (7.28) becomes

$$\dot{S}(t) = F(\mathbf{x}) + G(q)u + v - y_d^{(2)} + \eta(\mathbf{x}, t) \tag{7.29}$$

The following state feedback control law is assumed:

$$u = G^{-1}(q)[-F(\mathbf{x}) - v - \Lambda S] \tag{7.30}$$

where $\Lambda > 0$ is the feedback control gain matrix that is determined by users. It is not difficult to prove that the proposed controller (7.30) can achieve a uniform boundedness of errors.

7.5 Accommodation Control after Fault Detection

When the isolation decision is made, we need to reconfigure the controller. Without loss of generality, we assume that the rth fault is isolated. The accommodation controller is given by

$$u = u_n - G^{-1}(q)\bar{\theta}_r \zeta_r^f(t) \tag{7.31}$$

where u_n is the normal control (7.30). Substituting this control into (7.28) produces the same equation (7.29). The stability of the closed-loop system is guaranteed.

In some cases, the detected fault cannot be isolated. For example, a new fault is not classified into the fault function set Ω_f. In this case, the accommodation control based on the isolation decision cannot be used. Since the fault function is unknown, a neural network (NN) approximator is used to solve this problem. We assume that the fault function $\zeta(\mathbf{x})$ can be approximated by a general one-layer neural network as

$$\zeta(\mathbf{x}) = W^{*T}\Phi(\mathbf{x}) + \xi \tag{7.32}$$

where W^* is the ideal weight of NN and the bounded function approximation error ξ satisfies $\|\xi\| \le \xi_M$ with constant ξ_M. The NN approximation error ξ represents the minimum possible deviation between the unknown function and the function estimation. In general, increasing the NN node number reduces the error ξ. The ideal weights W^* are unknown and need to be estimated for controller design. Let $\hat{W}$ be estimates of the ideal W^*. Then, an estimate $\hat{\zeta}(\mathbf{x})$ of $\zeta(\mathbf{x})$ can be given by

$$\hat{\zeta}(\mathbf{x}) = \hat{W}^T\Phi(\mathbf{x}) \tag{7.33}$$

Therefore, the fault accommodation control law is reconfigured by

$$u = G^{-1}(q)[-F(\mathbf{x}) - v - \Lambda S - \hat{W}^T\Phi(\mathbf{x})] \tag{7.34}$$

with the learning law

$$\dot{W} = \Upsilon \Phi(\mathbf{x}) S^T - \rho \Upsilon (\hat{W} - W_a) \tag{7.35}$$

where $\Upsilon = \Upsilon^T > 0$, $\rho > 0$, and W_a is a design constant matrix. Substituting the control law (7.34) into the system (7.28), we have the following closed-loop system:

$$\dot{S}(t) = -\Lambda S + \eta(\mathbf{x}, t) - y_d^{(2)} + \mathcal{B}(t - T)\varsigma(\mathbf{x}) - \hat{W}^T \Phi(\mathbf{x})$$

The term $\mathcal{B}(t - T)\varsigma(\mathbf{x}) - \hat{W}^T \Phi(\mathbf{x})$ will satisfy the following relationship:

$$\mathcal{B}(t - T)\varsigma(\mathbf{x}) - \hat{W}^T \Phi(\mathbf{x}) = \tilde{W}^T \Phi(\mathbf{x}) - \Theta(t) W^{*T} \Phi(\mathbf{x}) + \mathcal{B}(t - T)\xi \tag{7.36}$$

where

$$\tilde{W} = W^* - \hat{W},$$

$$\Theta(t) = diag\{e^{-\theta_1(t-T)}, e^{-\theta_2(t-T)}, \ldots, e^{-\theta_m(t-T)}\}$$

Note that $\mathcal{B}(t - T)$ is bounded; therefore, $\|\mathcal{B}(t - T)\xi\| \leq \xi_M$. Utilizing these results and the Lyapunov theory, the system stability can be established.

7.6 Extension to Output Feedback Control Design

In the preceding section, all the results have been obtained under the assumption that the full state of the system can be measured. This assumption will be removed in this section and a more realistic problem where only a part of the states can be measured will be considered.

$F(\mathbf{x})$ is Lipschitz in $\mathbf{x}$, i.e.,

$$\|F(\mathbf{x}) - F(\hat{\mathbf{x}})\| \leq L_F \|\mathbf{x} - \hat{\mathbf{x}}\|$$

with constant L_F.

7.6.1 Fault Diagnosis Scheme

7.6.1.1 Fault Detection

First, the following observer is constructed:

$$\dot{\hat{\mathbf{x}}} = A_0\hat{\mathbf{x}} + L_0(y - \hat{y}) + b[F(\hat{\mathbf{x}}) + G(q)u] \tag{7.37}$$

$$\hat{y} = C\hat{\mathbf{x}} \tag{7.38}$$

Define the state and output estimation errors by $\tilde{\mathbf{x}} = \mathbf{x} - \hat{\mathbf{x}}$ and $\tilde{y} = y - \hat{y}$, respectively. It can be easily derived that the dynamics of residual generator is governed by

$$\dot{\tilde{x}} = \bar{A}\tilde{x} + b[F(x) - F(\hat{x})] + b\eta(x, t) \tag{7.39}$$

$$\tilde{y} = C\tilde{x} \tag{7.40}$$

where $\bar{A} = A_0 - L_0 C$. The gain matrix L_0 is chosen so that $\bar{A}$ is stable. The solution of the above equation is

$$\tilde{x} = e^{\bar{A}t}\tilde{x}(0) + \int_0^t e^{\bar{A}(t-\tau)} b[F(x) - F(\hat{x}) + \eta] d\tau \tag{7.41}$$

Note that there exist constants δ_1, β_1 such that

$$||e^{\bar{A}t}|| \le \delta_1 e^{-\beta_1 t}$$

Taking the norm to the above equation, it follows that

$$||\tilde{x}|| \le \delta_1 e^{-\beta_1 t}||\tilde{x}(0)|| + \int_0^t \delta_1 e^{-\beta_1(t-\tau)}||b||(\bar{\eta} + L_F||\tilde{x}||) d\tau$$

By applying Gronwall's lemma,

$$||\tilde{x}|| \le \kappa_1(t) + \delta_1||b||L_F \int_0^t \kappa_1(\tau) e^{-(\beta_1 - \delta_1||b||L_F)(t-\tau)} d\tau \tag{7.42}$$

where

$$\kappa_1(t) = \delta_1 e^{-\beta_1 t}||\tilde{x}(0)|| + \delta_1||b||\bar{\eta}\frac{1 - e^{-\beta_1 t}}{\beta_1}$$

From $\tilde{y} = C\tilde{x}$, the threshold is given by

$$||\tilde{y}|| \le ||C||\left[\kappa_1(t) + \delta_1||b||L_F \int_0^t \kappa_1(\tau) e^{-(\beta_1 - \delta_1||b||L_F)(t-\tau)} d\tau\right] = \varpi(t)$$

In the above result, $||\tilde{x}(0)||$ can be replaced by a conservative estimate ι_0, where $||\tilde{x}(0)| \le \iota_0$. The fault detection can be carried out as

$$\begin{cases} ||\tilde{y}|| \le \varpi(t), & \text{no fault occurs} \\ ||\tilde{y}|| > \varpi(t), & \text{fault has occurred} \end{cases}$$

where $\varpi(t)$ is the threshold of the fault detection.

7.6.1.2 *Fault Isolation*

For an isolating scheme, $s + 1$ isolation estimators are defined following the isolation algorithm presented previously:

$$\dot{\hat{x}}_r = A_0\hat{x}_r + L_0(y - \hat{y}) + b[F(\hat{x}_r) + G(q)u + \bar{\theta}_r \zeta_r^f]$$

$$\hat{y}_r = C\hat{x}_r, \ r \in [0, s]$$

Taking a similar procedure, we can obtain the threshold of the isolation:

$$\varpi_r(t) = ||C|| \left[\kappa_2(t) + \delta_1 ||b|| L_F \int_0^t \kappa_2(\tau) e^{-(\beta_1 - \delta_1 ||b|| L_F)(t-\tau)} d\tau \right]$$

where $\kappa_2(t) = \delta_1 e^{-\beta_1 t} ||\tilde{x}(0)|| + \delta_1 ||b|| [\bar{\eta} \frac{1-e^{-\beta_1 t}}{\beta_1} + \mathcal{O}_M \int_0^t e^{-\beta_1(t-\tau)} ||\zeta_r^f(\tau)|| d\tau]$.

The isolation is made when the estimation error $|\tilde{y}_r(t)| = |y - \hat{y}_r|$ is less than its corresponding threshold bound ϖ_r.

Similar to preceding section, the fault identification technique can be extended to the output feedback case.

7.6.2 Robust Control for the System under No-Fault Condition

For the design of fault accommodation control, a state error system has to be introduced. Define the state error

$$e_x = \mathbf{x} - \mathbf{x}_d$$

where

$$\mathbf{x}_d = [y_{d1}, \dot{y}_{d1}, y_{d2}, \dot{y}_{d2}, \ldots, y_{dm}, \dot{y}_{dm}]^T \in R^{2m}$$

System (7.3) may be expressed as

$$\dot{e}_x = A e_x + b \left[F(\mathbf{x}) + G(q)u - y_d^{(2)} + K e_x + \eta(\mathbf{x}, t) + \mathcal{B}(t - T)\zeta(\mathbf{x}) \right] \quad (7.43)$$

$$e_y = C e_x \quad (7.44)$$

where $A = A_0 - bK$, $y_d^{(2)} = [y_{d1}^{(2)}, \ldots, y_{dm}^{(2)}]^T$ and $e_y = y - y_d$. The constant matrix K in (7.43) is chosen so that $C[sI - A]^{-1}b$ is strictly positive real (SPR).

For the system under a no-fault condition, i.e., $\mathcal{B}(t-T)\zeta(\mathbf{x}) = 0$, the feedback control is required to allow the output vector y to track reference y_d. Since the state in the system (7.3) is not available, the following observer is proposed:

$$\dot{\hat{x}} = A_0 \hat{x} + L_0(y - \hat{y}) + b [F(\hat{x}) + G(q)u + \Xi_1] \quad (7.45)$$

where $\Xi_1 = (\frac{2}{\lambda_{min}(Q_1)} L_F^2 + \frac{1}{2})\hat{y}$. The gain matrix L is chosen so that $C[sI - (A_0 - L_0C)]^{-1}b$ is SPR. When (7.45) is subtracted from (7.3), the following observation error system is obtained:

$$\dot{\tilde{x}} = \bar{A}\tilde{x} + b[F(\mathbf{x}) - F(\hat{x}) + \eta - \Xi_1]$$

where $\bar{A} = A_0 - L_0C$. Using (7.45), the following controller is proposed for the system under a no-fault condition:

$$u = G^{-1}(q)u_0 \quad (7.46)$$

$$u_0 = -K_e e_y - F(\hat{x}) + y_d^{(2)} \quad (7.47)$$

where $K_e > 0$ with $\lambda_{min}(K_e) > \frac{2}{\lambda_{min}(Q_1)} L_F^2 + \frac{1}{\lambda_{min}(Q_2)} ||K||^2 + \frac{1}{2}$ (Q_1 and $Q_2 > 0$).

7.6.3 Accommodation Control after Fault Detection

When a fault is detected and the failure is isolated, the reconfigured controller is given by

$$u = G^{-1}(q)[u_0 - \bar{\theta}_r \zeta_r^f(t)] \tag{7.48}$$

The stability of the closed-loop system can be guaranteed under this control.

When the failure cannot be isolated, the accommodation in the control system can be achieved through adding an NN approximator into the normal controller. The proposed accommodation control is

$$u = G^{-1}(q)[u_0 - \hat{W}^T \Phi(\hat{x})] \tag{7.49}$$

with adaptive law $\dot{\hat{W}} = \Upsilon \Phi(\hat{x})(\tilde{y}^T + e_y^T) - \rho \Upsilon(\hat{W} - W_a)$. The reconfigured observer is given by

$$\dot{\hat{x}} = A_0 \hat{x} + L_0(y - \hat{y}) + b \left[F(\hat{x}) + G(q)u + \Xi_1 + \frac{1}{2}\tilde{y} + \hat{W}^T \Phi(\hat{x}) \right]$$

The resulting observation error equation is

$$\dot{\tilde{x}} = \bar{A}\tilde{x} + b \left[F(x) - F(\hat{x}) + \eta - \Xi_1 + \mathcal{B}(t - T)\zeta - \hat{W}^T \Phi(\hat{x}) - \frac{1}{2}\tilde{y} \right] \tag{7.50}$$

The proposed controller can achieve a stable system.

7.7 Experimental Tests

To illustrate the effectiveness of the proposed method, real-time experiments are carried out on a Cartesian robotic system (see Figure 7.3) manufactured by Anorad Co, Shirley, New York. The dSPACE control development and rapid prototyping system, in particular, the DS1103 board, is used.

The following model for X–Y-axes is considered:

$$\left. \begin{array}{l} \ddot{x}_1 = -a_1 \dot{x}_1 + b_1 u_1 - f_{fric1} + \eta_1(x, t) \\ \ddot{x}_2 = -a_2 \dot{x}_2 + b_2 u_2 - f_{fric2} + \eta_2(x, t) \end{array} \right\} \tag{7.51}$$

The dominant model of $\ddot{x} = -a\dot{x} + bu$; it is obtained from velocity response using a step input. Figure 7.4 shows the step responses of actual and identified models, respectively, where the solid line represents the actual response, while the dotted line represents the model response. The model error is due to the nonlinear terms in the system. The frictions f_{fric1}, f_{fric2} are determined by

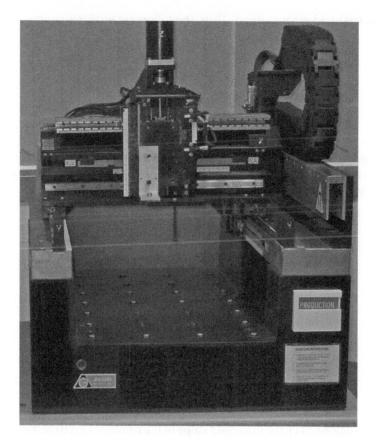

FIGURE 7.3
Control test bed.

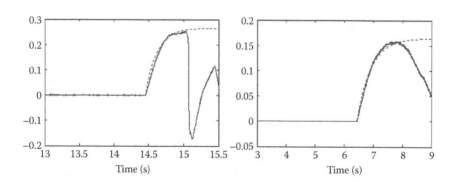

FIGURE 7.4
Actual and model responses: X-axis (left), Y-axis(right).

using a trial-and-error method. Thus, the following model is used:

$$\begin{aligned}
\ddot{x}_1 = \quad & -6.176\dot{x}_1 + 0.547u_1 - 0.164\,\text{sgn}(\dot{x}_1) \\
& -0.115e^{-\frac{\dot{x}_1^2}{0.001^2}}\,\text{sgn}(\dot{x}_1) + \eta_1(\mathbf{x}, t) \\
\ddot{x}_2 = \quad & -2.110\dot{x}_2 + 0.058u_2 - 0.166\,\text{sgn}(\dot{x}_2) \\
& -0.121e^{-\frac{\dot{x}_2^2}{0.001^2}}\,\text{sgn}(\dot{x}_2) + \eta_2(\mathbf{x}, t)
\end{aligned} \qquad (7.52)$$

The functions $\eta_1(\mathbf{x}, t)$ and $\eta_2(\mathbf{x}, t)$ are the remaining nonlinear uncertainties and are bounded by 2.6 and 1.2. The estimator used for residual generator is given by

$$\begin{aligned}
\dot{\hat{x}} = & \begin{bmatrix} 0 & 1 & 0 & 0 \\ 0 & 0 & 0 & 0 \\ 0 & 0 & 0 & 1 \\ 0 & 0 & 0 & 0 \end{bmatrix} \hat{x} \\
& + \begin{bmatrix} 0 & 0 \\ 1 & 0 \\ 0 & 0 \\ 0 & 1 \end{bmatrix} \begin{bmatrix} -6.176\dot{x}_1 + 0.547u_1 - \left(0.164 + 0.115e^{-\frac{\dot{x}_1^2}{0.001^2}}\right)\text{sgn}(\dot{x}_1) \\ -2.110\dot{x}_2 + 0.058u_2 - \left(0.166 + 0.121e^{-\frac{\dot{x}_2^2}{0.001^2}}\right)\text{sgn}(\dot{x}_2) \end{bmatrix} \\
& + \begin{bmatrix} 60 & 0 & 0 & 0 \\ 70 & 6.176 & 0 & 0 \\ 0 & 0 & 50 & 0 \\ 0 & 0 & 100 & 2.11 \end{bmatrix} \begin{bmatrix} \tilde{x}_1 \\ \dot{\tilde{x}}_1 \\ \tilde{x}_2 \\ \dot{\tilde{x}}_2 \end{bmatrix}
\end{aligned}$$

From Equation (7.17), the threshold value of the residual signal can be derived as

$$\varpi_1 = \frac{1}{51.1561} \int_0^t 2.6[e^{-7.5103(t-\tau)} - e^{-58.6664(t-\tau)}]d\tau \le 0.0020 \times 2.6 = 0.0052$$

$$\varpi_2 = \frac{1}{43.5137} \int_0^t 1.2[e^{-4.2982(t-\tau)} - e^{-47.8119(t-\tau)}]d\tau \le 0.0046 \times 1.2 = 0.0055$$

since the initial states are zero. In the experiment, the X-Y linear motors follow the trajectories $x_{1d} = 0.015\sin(w_1t)$ m, where $w_1 = 1$ rad/s and $x_{2d} = 0.02\sin(w_2t)$, where $w_2 = 2$ rad/s. The parameters in the control (7.30) are chosen as $k_1 = 100$, $k_2 = 10$, $\Lambda = \begin{bmatrix} 24 & 0 \\ 0 & 100 \end{bmatrix}$. Figure 7.5 shows the control responses and residuals under normal conditions. The residuals are oscillatory since the X-Y motors follow sinewave signals. It is observed that each residual is below its corresponding threshold and no fault is detected. Now, a mechanical fault is considered on the Y-axis due to obstruction from the cable protection chain, which consists of many chain links (see Figure 7.6). Due to prolonged high-speed operation, one or several of the chain links are jammed, which obstructs the motor movement. Figure 7.7 shows the response due to the

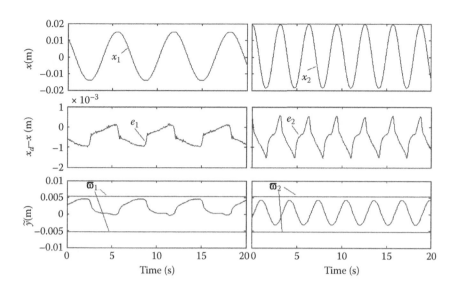

FIGURE 7.5
Tracking performance and residual signal under normal control.

fault occurrence. It is observed that the residual of the X-axis is still below its threshold, but the residual of the Y-axis is beyond its threshold. It should be noted that the tracking performance is degraded due to fault occurrence. In this situation, the fault isolation algorithm (7.20) with $\varsigma_2^f(t) = \begin{bmatrix} 0 \\ 0.4 + 3.1|x_2| \end{bmatrix}$ is used to find the characteristics of the failure. After that, the accommodation control (7.31) incorporating the isolation information is activated. Figure 7.8

Chain Pin   Chain Link

FIGURE 7.6
Cable protection chain.

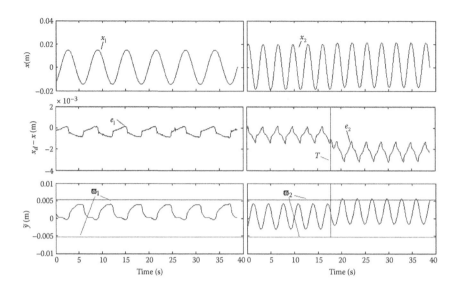

FIGURE 7.7
Tracking performance and residual signal under normal control when the fault occurs.

shows the control performance. It is observed that the fault is isolated successfully, and the tracking performance is improved immediately when applying the accommodation control.

It is possible that in some cases the detected fault cannot be isolated. In this situation, the NN accommodation control (7.34) will be triggered. For the purpose of illustrating the NN compensation, it is assumed that the mechanical fault above cannot be isolated and the NN control (7.34) is tested. Figures 7.9 and 7.10 show the control performance and adaptive NN compensation signal, respectively. It is observed that once the residual signal exceeds the threshold value, the NN accommodation control is triggered to compensate the effects of the fault. The results show that the control performance is improved following the adaptive learning.

7.8 Concluding Remarks

In this chapter, the fault diagnosis and accommodation schemes have been presented for a MIMO precise actuator system. The threshold of the residual generator has been derived so that the fault diagnosis can be made. The accommodation control schemes have been developed based on NN compensation. Experimental results have confirmed that the designed residual generator can monitor the operation of the machine and the reconfigured controller can improve the tracking performance against the system failure.

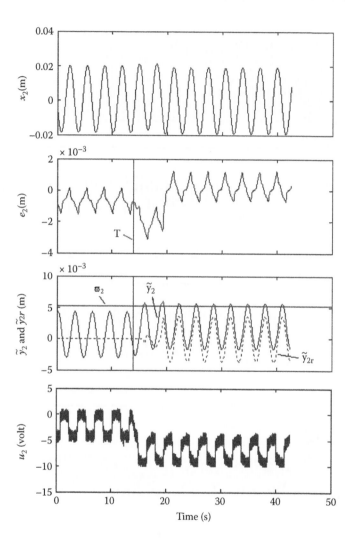

FIGURE 7.8
Tracking performance and residual signal under accommodation control (7.31).

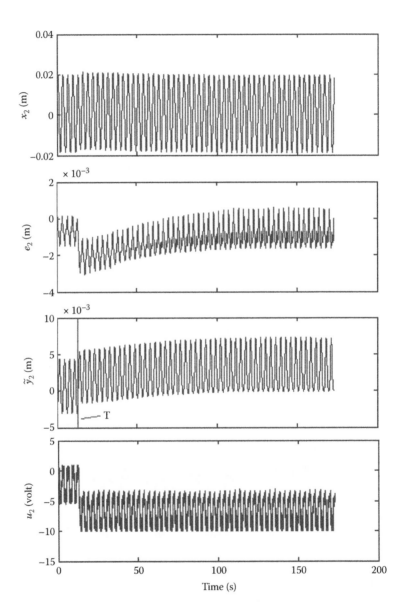

FIGURE 7.9
Tracking performance and residual signal under NN accommodation control.

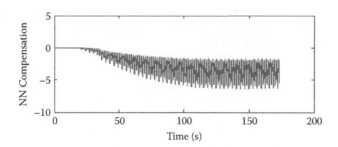

FIGURE 7.10
Adaptive NN output signal.

References

1. X. Q. Liu, H. Y. Zhang, J. Liu, and J. Yang. Fault detection and diagnosis of permanent-magnet DC motor based on parameter estimation and neural network. *IEEE Transactions on Industrial Electronics*, 47, 1021–1030, 2000.
2. O. Bhattacharyya, M. B. Jun, S. G. Kapoor, and R. E. DeVor. The effects of process faults and misalignments on the cutting force system and hole quality in reaming. *International Journal of Machine Tools and Manufacture*, 46, 1281–1290, 2006.
3. S. H. Kia, H. Henao, and G.-A. Capolino. A high-resolution frequency estimation method for three-phase induction machine fault detection. *IEEE Transactions on Industrial Electronics*, 54, 2305–2314, 2007.
4. S. N. Huang, K. K. Tan, Y. S. Wong, C. W. de Silva, H. L. Goh, and W. W. Tan. Tool wear detection and fault diagnosis based on cutting force monitoring. *International Journal of Machine Tools and Manufacture*, 47, 444–451, 2007.
5. S. N. Huang, K. K. Tan, and T. H. Lee. Sliding mode monitoring and control of linear drives. *IEEE Transactions on Industrial Electronics*, 56(9), 3532–3540, 2009.
6. S. N. Huang, K. K. Tan, and T. H. Lee. Automated fault detection and diagnosis in mechanical systems. *IEEE Transactions on SMC, Part C*, 37(6), 1360–1364, 2007.
7. S. N. Huang and K. K. Tan. Fault detection and diagnosis based on modeling and estimation methods. *IEEE Transactions on Neural Networks*, 20(5), 872–881, 2009.
8. S. Wu and T. W. S. Chow. Induction machine fault detection using SOM-based RBF neural networks. *IEEE Transactions on Industrial Electronics*, 51, 183–194, 2004.
9. W. W. Tan and H. Huo. A generic neurofuzzy model-based approach for detecting faults in induction motors. *IEEE Transactions on Industrial Electronics*, 52, 1420–1427, 2005.
10. M. S. Ballal, Z. J. Khan, H. M. Suryawanshi, and R. L. Sonolikar. Adaptive neural fuzzy inference system for the detection of inter-turn insulation and bearing wear faults in induction motor. *IEEE Transactions on Industrial Electronics*, 54, 250–258, 2007.
11. Y. Diao and K. M. Passino. Stable fault-tolerant adaptive fuzzy/neural control for a turbine engine. *IEEE Transactions on Control Systems Technology*, 9, 494–509, 2001.

12. M. M. Polycarpou and A. J. Helmicki. Automated fault detection and accommodation: A learning systems approach. *IEEE Transactions on Systems, Man, and Cybernetics*, 25, 1447–1458, 1995.

13. J. A. Farrell, T. Berger, and B. D. Appleby. Using learning techniques to accommodate unanticipated faults. *IEEE Control Systems Magazine*, 13, 40–49, 1993.

14. M. L. Visinsky, J. R. Cavallaro, and I. D. Walker. A dynamic fault tolerance framework for remote robots. *IEEE Transactions on Robotics and Automation*, 11, 477–490, 1995.

15. R. L. A. Ribeiro, C. B. Jacobina, E. R. C. da Silva, and A. M. N. Lima. Fault-tolerant voltage-fed PWM inverter AC motor drive systems. *IEEE Transactions on Industrial Electronics*, 51, 439–446, 2004.

16. R. Patton, P. Frank, and R. Clark. *Fault diagnosis in dynamic systems*. Upper Saddle River, NJ: Prentice-Hall, 1989.

17. Diesel Service and Supply. The many causes of power failures. http://www.dieselserviceandsupply.com/ Causes_of_Power_Failures.aspx.

18. V. I. Rudney. Systematic analysis of induction coil failures. *Heat Treating Progress*, September/October 2005.

8

Case Studies of Precise Actuator Applications

This chapter is devoted to three case studies involving the application of precise actuators. The main purpose of this chapter is to highlight the key issues involved in the practical application of motion control to real precise actuators and bring forth the motivation behind the work in preceding chapters.

The first case study describes a robust adaptive control method for positioning piezoelectric actuators (ultrasonic motor) to achieve highly precise motion. The model employed to describe the motor is a second-order linear model plus a nonlinear part comprising predominantly a dynamical hysteresis. Based on the model, the overall control algorithm uses a proportional-integral-derivative (PID) component and an adaptive robust component for estimating the parameters of the piezo motor model. The adaptive component is continuously refined based on just prevailing input and output signals. Real-time experimental results are provided to verify the effectiveness of the proposed scheme when applied to high-precision motion trajectory tracking such as intracytoplasmic sperm injection (ICSI).

The second case presents the use of a motion control stage to treat a common ear disease called otitis media with effusion (OME), involving the surgeon inserting a grommet in the eardrum to bypass the eustachian tube in draining fluid when medication fails. In this study, a device for myringotomy and grommet tube insertion is first designed and introduced. Due to the advantages of the high precision and fast response of the piezomotor, a 2-DOF (degrees of freedom) ultrasonic stage, which consists of two piezomotors, is chosen to provide the motion sequences of the device, especially precise path tracking during the grommet insertion. This study briefly presents the mechanical design of the device and the configuration and control of the 2-DOF stage. The model of the stage consisting of a linear and nonlinear term is built. A PID controller is used as the main controller and tuned with the help of the linear quadratic regulator (LQR) technique. Since there are nonlinear dynamics caused by friction and hysteresis existing in the system, a nonlinear compensation including a sign function and sliding mode control is designed to reject the nonlinearity. Moreover, a decoupling controller is designed to eliminate the coupling effects between the two piezomotors stages. The experimental results show that the LQR-assisted PID controller with compensation can achieve very good system performance, and the decoupling controller can further improve the performance.

The third case presents a CNC machine surveillance system based on a vision-assisted thermal monitoring approach. First, a calibrated camera will be used to detect and track the milling and the position data obtained will be transmitted to a host computer. Subsequently, a laser built-in thermometer will continuously read the temperature of the milling by following the milling based on the position data. Finally, the host computer will generate an alarm signal when the temperature exceeds a preset threshold.

8.1 Robust Adaptive Control of Piezoelectric Actuators with an Application to Intracytoplasmic Sperm Injection

Among the electric motor drives, the piezoelectric actuator (PA) is one drive that is becoming very popular in high-precision applications. The increasingly widespread industrial applications of the PA in various optical fiber alignments, mask alignments, and medical micromanipulators are self-evident testimonies of the effectiveness of the PA in these application domains. The main benefits of a PA include low thermal losses and, most importantly, the high precision and accuracy achievable consequent of the direct drive principle. One major source of uncertainties in PA control design is the hysteresis behavior, which yields a rate-independent lag and residual displacement near zero input, reducing the precision of the actuators. Due to the typical precision positioning requirements and low offset tolerance of PA applications, the control of these systems, under the influence of these uncertainties, is particularly challenging since conventional PID control usually does not suffice in these application domains to meet the stringent performance requirements.

Hysteresis is an input-output nonlinearity with effects of nonlocal memory; i.e., the output of the system depends not only on the instantaneous input, but also on the history of its operation. For eliminating or reducing the effects of the hysteresis nonlinearity, a model-based hysteresis compensation technique will form a crucial component of the PA control strategy. Based on the linear mapping between the driving charge and displacement of the PA, the charge-driven technique is proposed by Newcomb and Flinn [1]. In Hagood et al. [2], an active structure control is proposed that is based on a linear transfer function model. In Cole and Clark [3] and Vipperman and Clark [4], the least mean square (LMS) algorithm is used to adaptively compensate the nonlinear dynamics affecting the PA. The Preisach model is used to compensate for the tracking error of piezoelectric actuators as proposed by Ge and Jouaneh [5]. In Stepanenko and Su [6], the Duhem model-based intelligent controller is designed. In Choi et al. [8], the Maxwell slip model is used to develop a feedforward compensator for a piezoelectric actuator. In Cruz-Hernandez and Hayward [7], the hysteresis nonlinearity is treated as a constant phase lag, and accordingly, a phase lead model is used to design a compensator. In Santa et al. [9], a neural network model is established that is then used in the controller design.

In this study, we consider the design and realization of a robust adaptive control algorithm to reject the hysteresis phenomenon associated with general PAs. The controller comprises a PID feedback component and an adaptive component for hysteresis compensation. The adaptive component is continuously refined based on just prevailing input and output signals. It will be proven that the tracking error can asymptotically converge to zero. In addition, analytical quantification is given to illustrate the improvement of the system's transient performance. Real-time experimental results verify the effectiveness of the proposed scheme for high-precision motion trajectory tracking using the piezoelectric actuator.

An important application to be explored and illustrated in this chapter focuses on intracytoplasmic sperm injection (ICSI). ICSI is a human-assisted method for animal or human reproduction, which was first introduced by Palermo et al. [10], and it has since been used throughout the world [11]. However, the survival and fertilization rates of oocytes from the ICSI process are still widely varying among practicing hospitals and institutions. Reported survival rates have ranged from 80 to 90%, and fertilization rates of intact oocytes have ranged from 45 to 70%. The proposed piezoelectric actuator is used in ICSI and cell impalement. The main requirement is to employ an automated method to execute a highly precise piercing motion through a soft, elastic, movable ball membrane whose diameter is about 80–150 μm with almost no deformation and no damage using a needle whose diameter is about 10 μm. The results show that the piezo control system can achieve a high oocyte survival rate.

8.1.1 Modeling of the Piezoelectric Actuator

The mathematical model for a voltage-controllable PA system can be approximately described by the differential equation [13]

$$m\ddot{x} = -K_f \dot{x} - K_g x + K_e(u(t) - F) \tag{8.1}$$

where $u(t)$ is the time-varying motor terminal voltage, $x(t)$ is the piezo position, K_f is the damping coefficient produced by the motor, K_g is the mechanical stiffness, K_e is the input control coefficient, m is the effective mass, and F is the system nonlinear disturbance. This model is based on Newton's second law, which is used in most piezo dynamics (see [8,13,14] and Chapter 3 of [16]).

One of the known disturbances present in the dynamics of the PA is the hysteresis phenomenon. Figure 8.1 shows the real-time open-loop response of a PA manufactured by Physik Instrumente. Typically, the magnitude of this hysteresis can constitute about 10–15% of the motion range [13]. Hysteresis

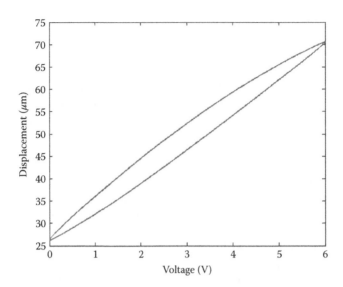

FIGURE 8.1
Hysteresis phenomenon.

generally impedes high-precision motion, and hysteresis analysis is thus a
key step toward the realization of high-performance PA systems. As elaborated in [12], the hysteresis is a friction-like phenomenon. In [14], a spring
coupled to a pure Coulomb friction element is used to model the piezo drive
with hysteresis. In [8], the Maxwell slip model is used to represent hysteresis.
However, these existing literatures use the static friction-like hysteresis models, which capture only part of the hysteresis phenomena. This is why [13]
presented a dynamical model for hysteresis, which is described by

$$\dot{z} = \alpha \dot{x} - \beta |\dot{x}| z - \gamma \dot{x} |z| \tag{8.2}$$

In this model, the hysteretic relationship between a state variable z and excitation x is introduced in the nonlinear mechanical vibrational analysis. The
parameter α controls the restoring force amplitude, and β and γ control the
shape of the hysteresis loop. However, the dynamical model in [13] is simple.
In [15], a more complex dynamical friction model from a frictional force was
presented. The model contains a state variable representing the average de-
flection of elastic bristles, which are a visualization of the topography of the
contacting surfaces. The resulting model shows most of the known friction
behavior, like hysteresis, friction lag, varying break-away forces, and stick-slip
motion. It is comprehensive enough to capture dynamical hysteresis effects.
Here, we employ this dynamical model:

$$F = \sigma_0 z + \sigma_1 \dot{z} + \sigma_2 \dot{x} \tag{8.3}$$

with

$$\dot{z} = \dot{x} - \frac{|\dot{x}|}{h(\dot{x})} z,$$

$$h(\dot{x}) = \frac{F_c + (F_s - F_c)e^{-(\dot{x}/\dot{x}_s)^2}}{\sigma_0}$$

where F_c, F_s, $\dot{x}_s$, σ_0, σ_1, and σ_2 are positive constants that are typically unknown.

Lemma 8.1 [15]
The internal state z is bounded.

The nonlinear function F can also be written as

$$F = (\sigma_1 + \sigma_2)\dot{x} + \sigma_0 z - \sigma_1 \frac{|\dot{x}|}{h(\dot{x})} z = (\sigma_1 + \sigma_2)\dot{x} + F_d(z, \dot{x}) \qquad (8.4)$$

The first part, $(\sigma_1 + \sigma_2)\dot{x}$, is a simple function of the velocity. The second part, $(\sigma_0 - \frac{\sigma_1|\dot{x}|}{h(\dot{x})})z$, is scaled by the z due to the dynamical perturbations in hysteresis. Since z and $h(\dot{x})$ are bounded,

$$|F_d(z, \dot{x})| = \left|\left(\sigma_0 - \sigma_1 \frac{|\dot{x}|}{h(\dot{x})}\right)z\right| \le k_1 + k_2|\dot{x}| \qquad (8.5)$$

where k_1 and k_2 are constants.

The above model allows us to design model-based control laws to directly compensate for the model parameters. In addition, a sliding mode control component can address the variable z effect and other unaccounted dynamics, such as the load force. The development of the robust adaptive control algorithm for estimating the parameters of the models will be presented in the next section.

8.1.2 Robust Adaptive Control

Consider the adaptive control problem of the system (8.1). Define the tracking error as $e = x_d - x$. The filtered tracking error is more commonly used in servo control, given by

$$r = K_I \int_0^t e(\tau)d\tau + K_p e + \dot{e} \qquad (8.6)$$

where K_I and $K_p > 0$ are chosen such that the polynomial $s^2 + K_p s + K_I$ is Hurwitz. Differentiating $r(t)$, (8.6) may be written in terms of the filtered tracking errors as

$$\frac{m}{K_e}\dot{r} = \frac{m}{K_e}(K_I e + K_p \dot{e} + \ddot{x}_d) + \frac{K_f}{K_e}\dot{x} + \frac{K_g}{K_e}x - (u - F)$$

$$= \frac{m}{K_e}(K_I e + K_p \dot{e} + \ddot{x}_d) + \left(\frac{K_f}{K_e} + \sigma_1 + \sigma_2\right)\dot{x} + \frac{K_g}{K_e}x - u + F_d \quad (8.7)$$

Define now the control input as

$$u = K_v r + \hat{a}_m(K_I e + K_p \dot{e} + \ddot{x}_d) + \hat{a}_{k\sigma}\dot{x} + \hat{a}_{ge}x + \hat{k}_1 sgn(r) + \hat{k}_2|\dot{x}|sgn(r) \quad (8.8)$$

where $K_v > 0$ is a constant, and $\hat{a}_m, \hat{a}_{k\sigma}, \hat{a}_{ge}, \hat{k}_1$, and $\hat{k}_2$ are estimates of $\frac{m}{K_e}, \frac{K_f}{K_e} + \sigma_1 + \sigma_2, \frac{K_g}{K_e}, k_1$ and k_2, respectively. Obviously, if $\hat{a}_m = \frac{m}{K_e}, \hat{a}_{k\sigma} = \frac{K_f}{K_e} + \sigma_1 + \sigma_2, \hat{a}_{ge} = \frac{K_g}{K_e}, \hat{k}_1 = k_1$, and $\hat{k}_2 = k_2$, then the control (8.8) leads to the closed-loop system expressed as

$$\frac{m}{K_e}\dot{r} = -K_v r - k_1 sgn(r) - k_2|\dot{x}|sgn(r) + F_d \quad (8.9)$$

For the Lyapunov function $L = \frac{m}{2K_e}r^2$ and using (8.5), $\dot{L} \leq -K_v r^2 - k_1|r| - k_2|\dot{x}||r| + k_1|r| + k_2|\dot{x}||r| = -K_v r^2$, it follows that $r \to 0$ as $t \to \infty$ since $K_v > 0$; i.e., the resulting system is asymptotically stable. Unfortunately, the hysteresis is unknown a priori in practice. In addition, it is also difficult to obtain the precise values of m, K_e, K_f, and K_g. However, the control law (8.8) suggests indeed that well-estimated functions can be used to improve the tracking performance. Motivated by this observation, an adaptive control technique, which continuously obtains and refines the model parameters, will next be developed.

Using the control law (8.8), we have

$$\frac{m}{K_e}\dot{r} = -K_v r + \tilde{a}_m(K_I e + K_p \dot{e} + \ddot{x}_d) + \tilde{a}_{k\sigma}\dot{x} + \tilde{a}_{ge}x$$

$$-\hat{k}_1 sgn(r) - \hat{k}_2|\dot{x}|sgn(r) + F_d \quad (8.10)$$

where $\tilde{a}_m = \frac{m}{K_e} - \hat{a}_m$, $\tilde{a}_{k\sigma} = \frac{K_f}{K_e} + \sigma_1 + \sigma_2 - \hat{a}_{k\sigma}$, $\tilde{a}_{ge} = \frac{K_g}{K_e} - \hat{a}_{ge}$.
We now specify the parameter update laws:

$$\dot{\hat{a}}_m = \gamma_1(K_I e + K_p \dot{e} + \ddot{x}_d)r \quad (8.11)$$

$$\dot{\hat{a}}_{k\sigma} = \gamma_2 \dot{x} r \quad (8.12)$$

$$\dot{\hat{a}}_{ge} = \gamma_3 x r \quad (8.13)$$

$$\dot{\hat{k}}_1 = \gamma_4|r| \quad (8.14)$$

$$\dot{\hat{k}}_2 = \gamma_5|\dot{x}||r| \quad (8.15)$$

where $\gamma_1, \gamma_2, \gamma_3, \gamma_4, \gamma_5 > 0$.

Theorem 8.1

Asymptotic convergence of robust adaptive controller.

   Consider the plant (8.1) and the control objective of tracking the desired trajectories, x_d, $\dot{x}_d$, $\ddot{x}_d$. The control law given by (8.8) with (8.11)–(8.15) ensures that the system states and parameters are uniformly bounded and that $r(t)$ asymptotically converges to zero.

Proof

We first define a Lyapunov function candidate $V(t)$ as

$$V(t) = \frac{1}{2}\frac{m}{K_e}r^2 + \frac{1}{2\gamma_1}\tilde{a}_m^2 + \frac{1}{2\gamma_2}\tilde{a}_{k\sigma}^2 + \frac{1}{2\gamma_3}\tilde{a}_{ge}^2 + \frac{1}{2\gamma_4}\tilde{k}_1^2 + \frac{1}{2\gamma_5}\tilde{k}_2^2 \quad (8.16)$$

where $\tilde{k}_1 = k_1 - \hat{k}_1$, $\tilde{k}_2 = k_2 - \hat{k}_2$. Taking the time derivative of V, it follows that

$$\dot{V} = \frac{m}{K_e}\dot{r}r + \frac{1}{\gamma_1}\tilde{a}_m\dot{\tilde{a}}_m + \frac{1}{\gamma_2}\tilde{a}_{k\sigma}\dot{\tilde{a}}_{k\sigma} + \frac{1}{\gamma_3}\tilde{a}_{ge}\dot{\tilde{a}}_{ge} + \frac{1}{\gamma_4}\tilde{k}_1\dot{\tilde{k}}_1 + \frac{1}{\gamma_5}\tilde{k}_2\dot{\tilde{k}}_2$$

$$= -K_vr^2 + [\tilde{a}_m(K_Ie + K_p\dot{e} + \ddot{x}_d) + \tilde{a}_{k\sigma}\dot{x} + \tilde{a}_{ge}x]r + [-\hat{k}_1sgn(r)$$

$$- \hat{k}_2|\dot{x}|sgn(r) + F_d]r + \frac{\tilde{a}_m\dot{\tilde{a}}_m}{\gamma_1} + \frac{\tilde{a}_{k\sigma}\dot{\tilde{a}}_{k\sigma}}{\gamma_2} + \frac{\tilde{a}_{ge}\dot{\tilde{a}}_{ge}}{\gamma_3} + \frac{\tilde{k}_1\dot{\tilde{k}}_1}{\gamma_4} + \frac{\tilde{k}_2\dot{\tilde{k}}_2}{\gamma_5}$$

Using the inequality (8.5) and $sgn(r)r = |r|$, it follows that

$$\dot{V} \leq -K_vr^2 + [\tilde{a}_m(K_Ie + K_p\dot{e} + \ddot{x}_d) + \tilde{a}_{k\sigma}\dot{x} + \tilde{a}_{ge}x]r - \hat{k}_1|r| - \hat{k}_2|\dot{x}||r|$$

$$+ k_1|r| + k_2|\dot{x}||r| + \frac{1}{\gamma_1}\tilde{a}_m\dot{\tilde{a}}_m + \frac{1}{\gamma_2}\tilde{a}_{k\sigma}\dot{\tilde{a}}_{k\sigma} + \frac{1}{\gamma_3}\tilde{a}_{ge}\dot{\tilde{a}}_{ge} + \frac{1}{\gamma_4}\tilde{k}_1\dot{\tilde{k}}_1 + \frac{1}{\gamma_5}\tilde{k}_2\dot{\tilde{k}}_2$$

$$= -K_vr^2 + [\tilde{a}_m(K_Ie + K_p\dot{e} + \ddot{x}_d) + \tilde{a}_{k\sigma}\dot{x} + \tilde{a}_{ge}x]r + [\tilde{k}_1|r| + \tilde{k}_2|\dot{x}||r|]$$

$$+ \frac{1}{\gamma_1}\tilde{a}_m\dot{\tilde{a}}_m + \frac{1}{\gamma_2}\tilde{a}_{k\sigma}\dot{\tilde{a}}_{k\sigma} + \frac{1}{\gamma_3}\tilde{a}_{ge}\dot{\tilde{a}}_{ge} + \frac{1}{\gamma_4}\tilde{k}_1\dot{\tilde{k}}_1 + \frac{1}{\gamma_5}\tilde{k}_2\dot{\tilde{k}}_2 \quad (8.17)$$

Substituting the expressions given by (8.11)–(8.15) yields

$$\dot{V} \leq -K_vr^2 \quad (8.18)$$

Since $K_v > 0$, it follows that $\dot{V} < 0$. This implies that r, $\hat{a}_m$, $\hat{a}_{k\sigma}$, $\hat{a}_{ge}$, $\hat{k}_1$, and $\hat{k}_2$ are uniformly bounded with respect to t. To show the boundedness of the tracking error r, we need to first prove that x and $\dot{x}$ are bounded. Define

$$\mu_0 = \int_0^t (x_d - x)d\tau \quad (8.19)$$

From (8.6), it follows that

$$
\begin{bmatrix} \dot{\mu}_0 \\ \ddot{\mu}_0 \end{bmatrix} = \begin{bmatrix} 0 & 1 \\ -K_I & -K_p \end{bmatrix} \begin{bmatrix} \mu_0 \\ \dot{\mu}_0 \end{bmatrix} + \begin{bmatrix} 0 \\ 1 \end{bmatrix} r \tag{8.20}
$$

Since K_I and K_p are chosen such that the polynomial $s^2 + K_I s + K_p$ is Hurwitz, the free system of the above equation is asymptotically stable. This observation, with r being bounded, implies that e is bounded. This also implies that $\dot{e}$ is bounded. Thus, x and $\dot{x}$ are bounded, since the reference signals x_d and $\dot{x}_d$ are bounded.

Since $r, \hat{a}_m, \hat{a}_{k\sigma}, \hat{a}_{ge}, \hat{k}_1$, and $\hat{k}_2$ are bounded and the inequality (8.5) holds, this implies that the right side of (8.10) is bounded. This further implies that $\dot{r}$ is bounded. Equation (8.18) and the positive definiteness of V then imply that

$$
\lim_{t \to \infty} \int_0^t -\dot{V}(\tau)d\tau = V(0) - \lim_{t \to \infty} V(t) < \infty \tag{8.21}
$$

By virtue of Barbalat's lemma, we have

$$
\lim_{t \to \infty} \dot{V}(t) = 0 \tag{8.22}
$$

Applying (8.18), it follows that

$$
\lim_{t \to \infty} r = 0 \tag{8.23}
$$

The proof is completed.

Theorem 8.1 relates only to the asymptotic performance of the signals in the closed-loop system; no transient performance is discussed. In practical applications, the transient performance is often more important. It is useful for tuning the parameters during the initial control phase. To this end, we have the following theorem.

Theorem 8.2
Transient Performance of Robust Adaptive Controller
 For the closed-loop system (8.1), (8.8), (8.11)–(8.15), the L_2 tracking error bound is

$$
||r||_2 \le \sqrt{\frac{\frac{m}{K_e}r^2(0) + \frac{1}{\gamma_1}\tilde{a}_m^2(0) + \frac{1}{\gamma_2}\tilde{a}_{k\sigma}^2(0) + \frac{1}{\gamma_3}\tilde{a}_{ge}^2(0) + \frac{1}{\gamma_4}\tilde{k}_1^2(0) + \frac{1}{\gamma_5}\tilde{k}_2^2(0)}{2K_v}}
$$

Proof
From (8.18), it follows that

$$
\dot{V} \le -K_v r^2 \tag{8.24}
$$

This can be written as

$$
K_v r^2 \le -\dot{V} \tag{8.25}
$$

Since $V = \frac{1}{2}\frac{m}{K_e}r^2 + \frac{1}{2\gamma_1}\tilde{a}_m^2 + \frac{1}{2\gamma_2}\tilde{a}_{k\sigma}^2 + \frac{1}{2\gamma_3}\tilde{a}_{ge}^2 + \frac{1}{2\gamma_4}\tilde{k}_1^2 + \frac{1}{2\gamma_5}\tilde{k}_2^2$ is nonincreasing and bounded from below by zero, it will be limited as $t \to \infty$, so that

$$\|r\|_2^2 = \int_0^\infty |r(\tau)|^2 d\tau \leq -\frac{1}{K_v}\int_0^\infty \dot{V}(\tau)d\tau$$

$$= \frac{1}{K_v}[V(0) - V(\infty)] \leq \frac{1}{K_v}V(0)$$

$$= \frac{1}{K_v}\left[\frac{m}{2K_e}r^2(0) + \frac{1}{2\gamma_1}\tilde{a}_m^2(0) + \frac{1}{2\gamma_2}\tilde{a}_{k\sigma}^2(0) + \frac{1}{2\gamma_3}\tilde{a}_{ge}^2(0)\right.$$

$$\left. + \frac{1}{2\gamma_4}\tilde{k}_1^2(0) + \frac{1}{2\gamma_5}\tilde{k}_2^2(0)\right]$$

which implies the conclusion held.

Remark 8.1: There is a sliding mode used in the controller. It is well known that a large gain will induce chattering in the presence of fast dynamics. To solve this problem, two ways are given: one is to use a very small initial $\hat{k}_1(\hat{k}_2)$ and adaptation gain $\gamma_4(\gamma_5)$; the other is to use a modified sliding mode $sat(r)$ as in [19]. For the parameters $\hat{k}_1, \hat{k}_2$, during the initial learning phase, the smaller initial values and adaptation gains should be chosen so as to learn the unknown model parameters and disturbance. After the initial phase, we can switch off the adaptation gains to prevent $\hat{k}_1, \hat{k}_2$ from increasing to infinity. In addition, a σ-modification term [20] can be added into the adaptive law to prevent the parameters $\hat{k}_1, \hat{k}_2$ increasing to infinity. For the parameters $\hat{k}_1, \hat{k}_2$, from a theoretical viewpoint, their derivative will approach zero as $t \to \infty$ since $r \to 0$ as $t \to \infty$. This implies that $\hat{k}_1, \hat{k}_2$ will eventually converge to stable values as $t \to \infty$. However, for a practical system, we propose that to learn the parameters $\hat{k}_1, \hat{k}_2$, the smaller initial values and adaptation gains should be chosen to avoid possible increasing in $\hat{k}_1, \hat{k}_2$ to infinity. Furthermore, the projection rule or σ-modification term (see [20]) can be added into the adaptive law to prevent the parameters $\hat{k}_1, \hat{k}_2$ increasing to infinity.

8.1.3 Experimental Results

In this subsection, we will illustrate the effectiveness of the proposed control scheme using real-time experiments. Figure 8.2 shows the experimental setup. The functional block diagram is shown in Figure 8.3. The PA used is a direct servo motor manufactured by Physik Instrumente, which has a travel length of 80 μm and is equipped with a linear variable differential transformer (LVDT) sensor with an effective resolution of 5 nm. The dSPACE control development and rapid prototyping platform is used. dSPACE integrated the entire development cycle seamlessly into a single environment, so that individual development stages between simulation and test can be run and rerun, without frequent readjustment. MATLAB®/Simulink can be used from

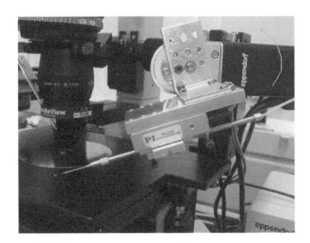

FIGURE 8.2
Experimental setup.

within the dSPACE environment. The entire control system is written into a Simulink object as shown in Figure 8.4, where the block "trajectory" is used to generate the desired trajectories and the block "controller1" is the adaptive controller itself.

To apply the controller, the system model (from control voltage to displacement) is first identified. The input-output signals are used as shown in Figures 8.5 and 8.6, where the frequency changes from 0.01 Hz to 200 Hz. The dominant linear model is computed to be

$$\ddot{x} = -1081.6\dot{x} - 5.9785 \times 10^5 x + 4.2931 \times 10^6 u \qquad (8.26)$$

Errors of the estimated model and actual piezo actuator are given in Figure 8.7. Clearly, the model error is caused by the nonlinear hysteresis, especially

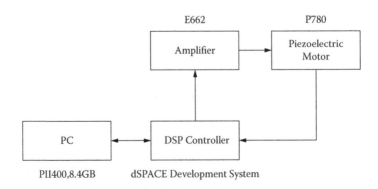

FIGURE 8.3
Block diagram of the experimental linear drive.

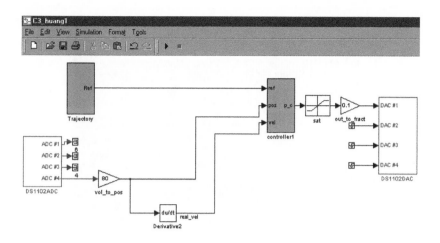

FIGURE 8.4
Software design for Simulink model.

in very high frequency. For our control application, the piezo is not worked in a very high frequency. Normally, it is only within 0–100 Hz. Thus, the second-order dominant model is still available.

The adaptive controller (8.8) is designed and applied to the piezoelectric actuator. The parameters of the controller are selected as

$$K_v = 0.00001, \; K_I = 400000, \; K_p = 100 \tag{8.27}$$

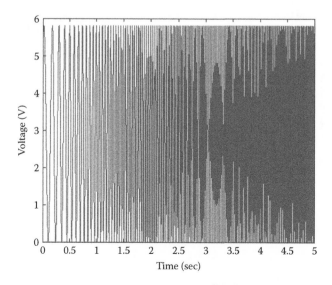

FIGURE 8.5
Input signal for identification.

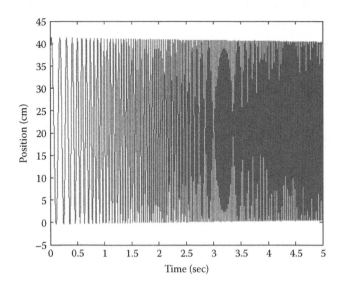

FIGURE 8.6
Output signal for identification.

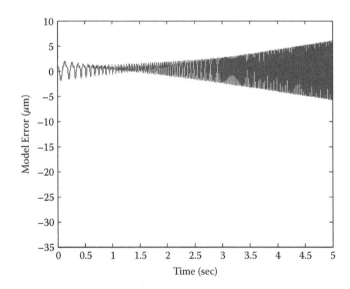

FIGURE 8.7
Errors between actual and estimated models.

As discussed in Remark 1, the transient performance can be improved by selecting the initial values as close to the actual ones as possible. Thus, the initial values for $\hat{a}_m$, $\hat{a}_{k\sigma}$ and $\hat{a}_{ge}$ can be chosen based on the identified model (8.26). They are $\hat{a}_m(0) = 2.3293 \times 10^{-7}$, $\hat{a}_{k\sigma}(0) = 2.5194 \times 10^{-4}$, and $\hat{a}_{ge}(0) = 0.1184$. The initial values for $\hat{k}_1$ and $\hat{k}_2$ are chosen as 10^{-7} and 10^{-8}, respectively. Since the mechanical structure and other components in the system have inherent unmodeled high-frequency dynamics that should not be excited, small adaptation factors are used, where we choose $\gamma_1 = \gamma_2 = \gamma_3 = 10^{-22}$, $\gamma_4 = \gamma_5 = 10^{-20}$. As only the displacement measurement is available in the linear drive system, the velocity is derived using a numerical difference method.

The first set of results pertains to tracking of pulse trains and the performance is shown in Figure 8.8. The parameters updated by the adaptive laws are shown in Figure 8.9. It is observed that a small micro-level accuracy can be achieved using the proposed adaptive control, and additionally, the response is fast, indicating a satisfactory transient performance. If only PID control in the proposed controller is used, the result is shown in Figure 8.10. It is observed that the adaptive controller can achieve faster response than that of the PID control.

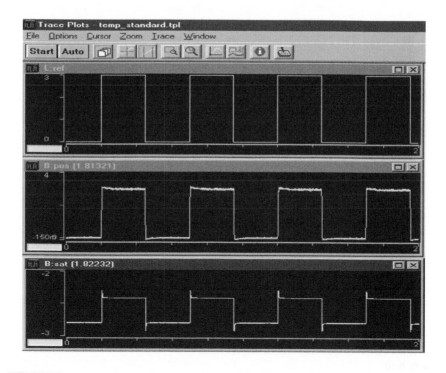

FIGURE 8.8
Square-wave responses with the control scheme: reference trajectory (μm) (top), actual response (μm) (middle), and control signal (V) (bottom).

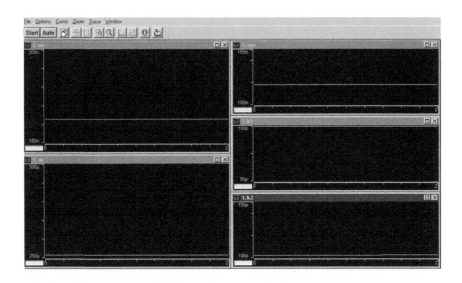

FIGURE 8.9

Model parameters with the control scheme: $am = \hat{a}_m, ak = \hat{a}_{k\sigma}, age = \hat{a}_{ge}, k1 = \hat{k}_1, k2 = \hat{k}_2$.

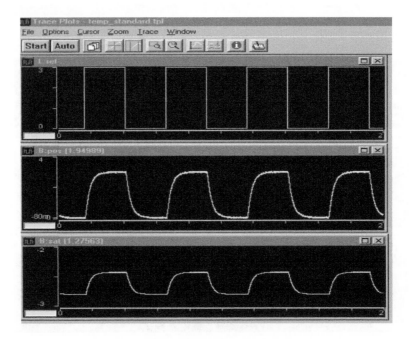

FIGURE 8.10

Square-wave responses with the control scheme (PID): reference trajectory (μm) (top), actual response (μm) (middle), and control signal (V) (bottom).

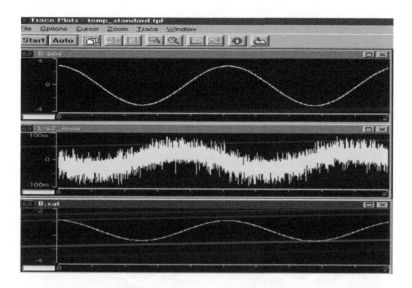

FIGURE 8.11
Sinewave responses with the control scheme: actual response (μm) (top), tracking error (μm) (middle), and control signal (V) (bottom).

The second set of results pertains to tracking of sinusoidal trajectories $A \sin(wt)$ where $A = 3$ μm, $w = 6$ rad/s. The control performance is shown in Figure 8.11. It can be observed that, under the proposed control, the actual response to the sinusoidal trajectories is good. The tracking error is about $\pm 100 \times 10^{-3}$ μm. If only PID control in the proposed controller is used, the tracking error is about $\pm 700 \times 10^{-3}$ μm, as shown in Figure 8.12. This shows that the adaptive controller can achieve better tracking performance than that of PID control.

8.1.4 Biomedical Application

Finally, in this section, the PA control system is put to real application in intracytoplasmic sperm injection (ICSI). ICSI has wide clinical applications. In order to achieve good results, it is important to restrict the possibility of oocyte injury as much as possible. This can be achieved with a fine and highly precise motion system. For this purpose, we employ a micromanipulator with PA under the proposed control scheme.

8.1.4.1 Instruments

An inverted microscope (Olympus, Tokyo, Japan) equipped with Hoffman modulation optics was used. For piezo micromanipulation, a Physik Instrumente piezo actuator unit (Germany) was attached to the hydraulic manipulator manufactured by Eppendorf (Germany). The configuration of the piezo

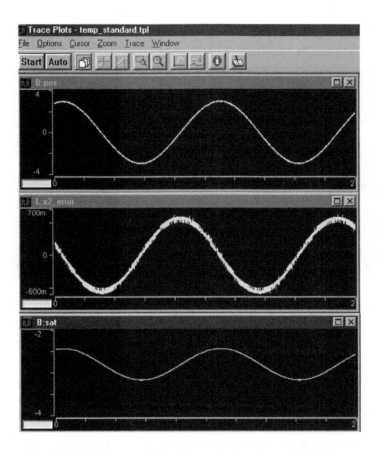

FIGURE 8.12
Sinewave responses with the control scheme (PID): actual response (μm) (top), tracking error (μm) (middle), and control signal (V) (bottom).

control system is shown in Figure 8.3. A finger switch is used to activate the actuator (Figure 8.13). The injection pipette is attached to the piezo actuator. The entire system is shown in Figure 8.14.

8.1.4.2 The Structure of Oocytes

From the engineering viewpoint, the oocyte is composed of three parts: the zona pellucida, the cytoplasm or vitelline, and the vitelline membrane or oolemma (Figure 8.15). The zona pellucida is a thick transparent membrane surrounding the ovum. The vitelline membrane, which is located between the vitelline and zona pellucida, is a protective membrane formed around the cytoplasm, acting as an obstruction to the entry of sperm. Usually, the zona pellucida is very elastic, and it can be difficult to be pierced at low speed. The main objective of ICSI is to pierce a needle, holding the sperm, through the zona pellucida and oolemma, and release the sperm in the deep area of

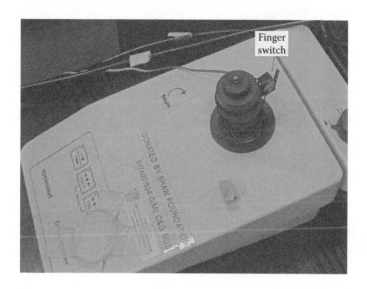

FIGURE 8.13
Finger switch.

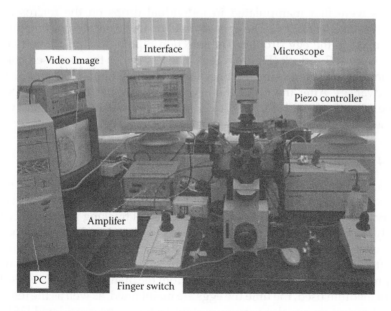

FIGURE 8.14
Photograph of the ICSI system.

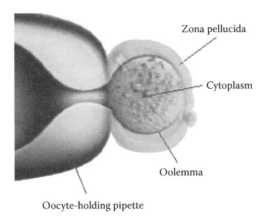

Zona pellucida

Cytoplasm

Oolemma

Oocyte-holding pipette

FIGURE 8.15
Photograph of an oocyte.

the cytoplasm. Minimum deformation of the oocyte should be achieved in order to maximize the survival of the egg after the artificial ICSI process.

8.1.4.3 ICSI Process

Oocytes were first manipulated in individual drops on the microscope's warming stage set at 37°C. The egg was held on to the oocyte-holding pipette and moved to the egg droplet. The injection pipette was positioned at the zona's edge at 3 o'clock (Figure 8.16a). The PA, under the proposed control scheme, was activated using the finger switch using rapid and stepwise (2–5 μm) advances. After the tip of the injection pipette has reached the perivitelline zone (Figure 8.16b), the PA was switched off using the finger switch and the pipette was withdrawn manually (Figure 8.16c).

8.1.4.4 Results

Table 8.1 lists the results of piezo micromanipulation on the oocytes from mice. The results relate to the survival of the oocytes after injection, an indirect beneficiary of the precise control performance. Actual tracking performance has already been illustrated in the previous section. One hundred sixty oocytes were collected for the piezo-ICSI. The oocytes were examined between 3 and 5 hs after injection. It should be pointed out, however, that the success of an ICSI process involves other factors, apart from the manipulation systems. The kind of medium used, the time the eggs are maintained, as well as the temperature of operation are also very important factors to be considered [18]. An average number of survival rates is 93.22%. This is higher than the conventional result (80–90%). In this application domain, it should be pointed out that a three percentage point improvement represents rather significant progress.

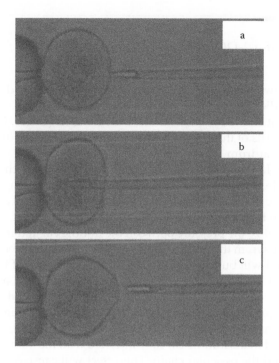

FIGURE 8.16

Photograph of piezo-ICSI. Photographs were prepared from video tape. (a) Piezo-ICSI: The tip of the injection pipette was positioned at the outer edge of the zona. (b) Then the injection needle penetrated the zona using piezo motor. (c) The needle was withdrawn slowly from the egg.

The technique reported here offered several advantages. First, there is no special requirement on the manufacturing of the needle. We do not require a fine needle tip. Second, with the control scheme, we do not require the use of toxic mercury in the pipette [17]. Third, the air syringe system is used, while the previous results [17] in piezo-ICSI design are based on an oil or water syringe system. This implies that the proposed system is easier to maintain. Finally, a high oocyte survival rate can be achieved.

TABLE 8.1

Results of Piezo Micromanipulation

Date of Experiments	Total No. Oocytes Injected	No. (%) Oocytes Survived
January 6, 2004	29	89.65
January 8, 2004	30	96.67
January 12, 2004	36	88.89
January 20, 2004	27	88.89
January 27, 2004	21	95.24
February 3, 2004	17	100

8.2 Control of a 2-DOF Ultrasonic Piezomotor Stage for Grommet Insertion

In this section, two ultrasonic piezomotors (USMs) are used to design a 2-DOF stage for a medical application involving myringotomy and grommet tube insertion for patients with otitis media with effusion (OME). Generally, the human ear consists of three main parts: the outer ear, the middle ear, and the inner ear. The eardrum (tympanic membrane) separates the ear canal from the middle ear. In the middle ear, there are three tiny bones that are used to transmit sound vibrations. One of the bones, named malleus, is attached to the inner surface of the eardrum. Behind the eardrum, a small duct called the eustachian tube links the middle ear to the back of the throat. The key functions are to equalize the pressure between the middle ear and the atmosphere and drain mucus from the middle ear. Once the eustachian tube is obstructed or blocked, fluid will build up inside the middle ear space. If the fluid is accumulated for a long time and cannot be drained, OME will arise, which is a very common ear disease affecting both children and adults worldwide. Normally, the fluid contains bacteria so that the ear is often infected, which can persist for many months each time in chronic OME. Moreover, the accumulation of fluid in the middle ear may interfere with the normal vibration of the eardrum and the ossicular chain. This effect will cause conductive hearing impairment. OME also causes body imbalance, discomfort, and reduces quality of life [21]. In more serious cases, OME may even result in irreversible damage to the middle ear structures, further complicating the treatment regimen.

Currently, the usual treatment of OME is to surgically insert an ear ventilation tube (also called grommet) in the eardrum to bypass the eustachian tube if medication fails. The tube will let the middle ear connect to the atmosphere again so as to allow the accumulated fluid to be easily drained out and evaporated. Generally, there are three main procedures in this surgery: (1) general anesthesia to keep the patient completely still during surgery, (2) myringotomy (under light microscopy visualization using a surgical knife) for making an incision onto the 8–10 mm diameter eardrum, and (3) grommet tube insertion: insert a short grommet tube carefully using micro-forceps and a needle in the incision. However, this conventional method has several disadvantages: (1) general anesthesia is associated with intubation trauma and respiratory and cardiac risks; (2) high surgical skills of surgeons are required; (3) costly operating theater time, equipment like surgical medical-grade microscopes, and theater set up with anesthetic and surgical assistants and nurses; (4) reduced access for patients in underserved areas of the world where proper hospitalization, general anesthesia, and well-equipped and skilled surgery are unavailable; and (5) delay in treatment during the wait and preparation for a formal surgery under general anesthesia.

To overcome the drawbacks of the conventional surgery for OME, an all-in-one handheld device allowing office-based grommet tube insertion in an awake patient for chronic OME is designed. The thickness of the eardrum can be expected to fall in the range of 30–120 μm, and it should be noted that the thickness distribution is not uniform across the membrane [22,23]. The ear canal is approximately 35 mm long and 5–10 mm in diameter. Normally, the space for myringotomy and grommet insertion in the eardrum is about a quarter of the eardrum, which is approximately 6.5–8 mm². Meanwhile, it is very important to avoid hurting the malleus during the surgery. Thus, highly precise motion is required to work within a small space. This is the main reason why USM is used in this device. The key procedures involving the USM are motion sequences to realize myringotomy and grommet insertion. The proposed motions required in both procedures are along two axis directions. Significantly, the motion is independent along each axis during myringotomy, while they are required to be cooperative during grommet insertion.

For the single-axis USM stage, the USM is driven based on the friction, so nonlinearities such as friction and hysteresis exist in the system. For the two-axis USM stage, the coupling phenomenon will appear if the installation between the two motors is not exactly orthogonal, or the center-of-gravity position of the load is not located at the center of the stage. In this work, to achieve a high-performance 2-DOF USM system for grommet insertion, two PID controllers with decoupling control for the 2-DOF stage are designed based on an LQR-assisted PID tuning method. A nonlinear compensation for the stage is designed by using a sign function and sliding mode control. In the following subsections, the background, including mechanical design of the device, control objective, and system modeling, is first presented. Then, the controller for the 2-DOF stage is designed. After that, some experiments are implemented to examine the control performance.

8.2.1 Background

The device for OME is shown in Figure 8.17. It consists of the following components:

1. A 2-DOF ultrasonic piezomotor stage that provides the required motion for myringotomy and grommet insertion along the Z-axis and the X-axis

2. A needle-shaped cutter used for making the incision in the eardrum and holding the grommet

3. A hollow holder used for locking the grommet and pushing the grommet into the incision

4. A servo motor with a link mechanism used for needle retraction

5. The sensing system, including a fiberscope camera system and force sensor

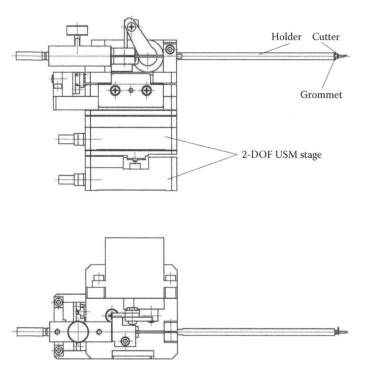

FIGURE 8.17
Mechanical design of the device for OME.

In this device, the USM stage is the key component. The 2-DOF stage is constructed with two M-663 linear single-axis USM stages manufactured by Physik Instrumente, as shown in Figure 8.18. One USM stage is mounted on the other stage orthogonally, so that the effect of coupling is minimal. The single-axis USM stage is mainly composed of a mover, stator, linear guideway, and built-in linear encoder with a resolution of 0.1 μm. The movement of this USM stage is generated by the friction between the piezo-ceramic plate fixed in the stator and the friction bar mounted on the mover. When a movement is required, the piezo-ceramic is excited by a specific electric field and produces high-frequency eigenmode oscillations. Therefore, the plate's tip moves along the guideway at that frequency and drives the mover forward or backward through the contact between the tip and the friction bar.

The specifications of the USM are shown as follows: the minimum incremental displacement of the mover is 0.3 μm, the maximum push–pull force is 2 N, and the velocity can be up to 400 mm/s. Moreover, a drive C-185 manufactured by Physik Instrumente is used to convert analog input signals (0 to ±10 V) into the required electric field and produce the required high-frequency oscillations. The analog input signals to the drive relate to the speed of the USM.

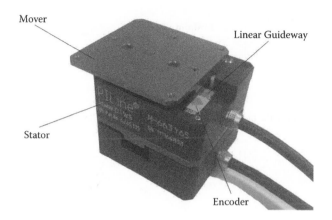

FIGURE 8.18
Ultrasonic piezomotor stage.

8.2.1.1 Control Objectives

In this chapter, the focus is on the control of the grommet insertion motion by using 2-DOF USM stage. A Shah-type grommet (pediatric ventilation tube) made of fluoroplastic is used, as shown in Figure 8.19. Its length is about 1.6 mm, and the outer, flange, and inner diameters are 1.2, 1.6, and 0.76 mm, respectively. In addition, there is a "tail" at one end of the grommet for inserting the grommet more easily.

During the grommet insertion procedure, as shown in Figure 8.20, the needle is retracted a bit into the holder first after the myringotomy. Then the holder associated with the needle guides the grommet move along the X-axis to the required position. Following that, the grommet is slid into the eardrum along a designed path that is a quarter cycle with a radius of 0.6 mm. Thus, the grommet's tail has been inserted in the eardrum when the holder reaches to the end of the path, as shown in Figure 8.20(e). Finally, two short moves along the X-axis and the Z-axis in sequence are applied so that the grommet can be inserted in the eardrum completely.

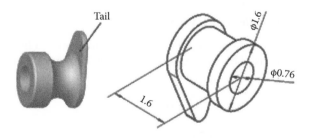

FIGURE 8.19
Shah-type grommet.

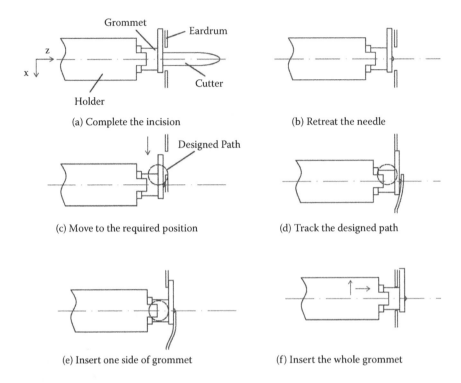

(a) Complete the incision   (b) Retreat the needle

(c) Move to the required position   (d) Track the designed path

(e) Insert one side of grommet   (f) Insert the whole grommet

FIGURE 8.20
Grommet insertion procedure.

Furthermore, the eardrum is slightly deformed during the grommet insertion procedure, as can be seen in Figure 8.20(d, e). In order to minimize the deformation and the damage on the eardrum, the insertion time should be as short as possible. Therefore, the control objective in this chapter is to achieve highly precise and fast 2-DOF motion for tracking the desired path in the grommet insertion procedure by using a 2-DOF USM stage.

8.2.2 System Modeling

8.2.2.1 Model of Single-Axis USM Stage

For a single-axis USM stage, the model is a combination of a linear term and a nonlinear term caused by friction and hysteresis. The model can be written as

$$\ddot{x}(t) = u_{linear}(t) + u_{nonlinear}(t) \tag{8.28}$$

where $x(t)$ is the position of the mover, and $u_{linear}(t)$ and $u_{nonlinear}(t)$ represent the linear term and the nonlinear term, respectively. For the USM, the linear term is the dominant part of the system.

In (8.28), the linear term can be considered a second-order system like the DC motor since the velocity output of the USM is obtained while the analog input is applied to the drive. Therefore, the linear term can be written as

$$u_{linear}(t) = -a\dot{x}(t) - bx(t) + cu(t) \tag{8.29}$$

where a, b, and c are the dominant parameters of the USM and $u(t)$ is the input signal to the drive.

The nonlinear dynamics includes hysteresis and friction, which relate to the velocity of the USM. Additionally, friction is the major nonlinear part of this USM. In this chapter, a Coulomb friction model $f_c(\dot{x})$ with a uncertain nonlinear component $\Delta f(\dot{x})$ is used to describe the nonlinear term. Therefore, the nonlinear term can be written as

$$u_{nonlinear}(t) = -f(\dot{x}) = -f_c(\dot{x}) - \Delta f(\dot{x}) \tag{8.30}$$

where $\Delta f(\dot{x})$ may be unknown but bounded, i.e., $|\Delta f(\dot{x})| \leq \Delta f_M$.

Hence, the model of the single-axis USM is obtained by combining (8.28) and (8.30), which is shown in (8.31).

$$\ddot{x}(t) = -a\dot{x}(t) - bx(t) + cu(t) - f_c(\dot{x}) - \Delta f(\dot{x}) \tag{8.31}$$

8.2.2.2 Model of 2-DOF USM Stage

For a 2-DOF USM stage, we consider the following dominant linear model:

$$\begin{bmatrix} X_1(s) \\ X_2(s) \end{bmatrix} = \begin{bmatrix} G_{11}(s) & G_{12}(s) \\ G_{21}(s) & G_{22}(s) \end{bmatrix} \begin{bmatrix} U_1(s) \\ U_2(s) \end{bmatrix} \tag{8.32}$$

where $X_1(s)$ and $X_2(s)$ represent the position outputs of the Z-axis USM stage and X-axis USM stage, and $U_1(s)$ and $U_2(s)$ represent the analog inputs to the Z-axis USM stage and X-axis USM stage, respectively.

To identify the model, a multifrequency square wave is chosen as the input signal. The square wave consists of three different frequencies, 10, 20, and 30 Hz, and an amplitude of 5 V.

First, we apply the square-wave input signal $u_1(t)$, while $u_2(t)$ is set to zero. In this case, the following equations are given by

$$X_1(s) = G_{11}(s)U_1(s)$$
$$X_2(s) = G_{21}(s)U_1(s) \tag{8.33}$$

The position outputs of both USM stages are shown in Figure 8.21. From the figure, it is found that $x_2(t)$ is affected by the Z-axis movement.

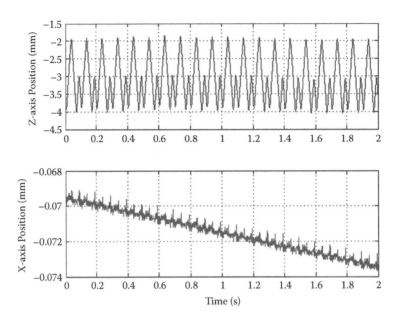

FIGURE 8.21
Position outputs to the input u_1.

Subsequently, let $u_2(t)$ be the square-wave input signal and $u_1(t)$ be zero. In this case, the following equations are given by

$$X_1 = G_{12}(s)U_2$$
$$X_2 = G_{22}(s)U_2 \tag{8.34}$$

The position outputs of both USM stages are shown in Figure 8.22. It is observed from the figure that there is almost no effect on the position along the Z-axis. This implies that $G_{12}(s) = 0$.

Combining (8.33) and (8.34), and assuming that the effects on X-axis motion by $u_1(t)$ behave as a second-order system, we have

$$X_1(s) = \frac{c_1}{s^2 + a_1 s + b_1} U_1(s) \tag{8.35}$$

$$X_2(s) = \frac{c_2}{s^2 + a_2 s + b_2} U_2(s) + \frac{c_3}{s^2 + a_3 s + b_3} U_1(s) \tag{8.36}$$

From (8.32), (8.35), and (8.36), we obtain the dominant two-axis linear model, which can be thought to be coupled since $G_{21}(s)$ does not equal zero.

$$\begin{bmatrix} X_1(s) \\ X_2(s) \end{bmatrix} = \begin{bmatrix} G_{11}(s) & 0 \\ G_{21}(s) & G_{22}(s) \end{bmatrix} \begin{bmatrix} U_1(s) \\ U_2(s) \end{bmatrix} \tag{8.37}$$

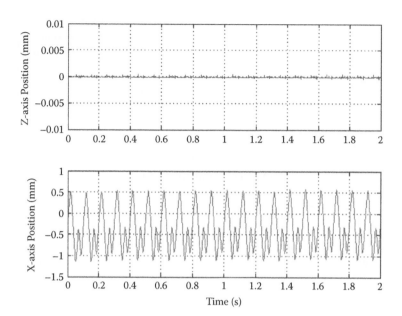

FIGURE 8.22
Position outputs to the input u_2.

Since the nonlinear term of the USM stage only affects itself, the complete model of the 2-DOF stage is given as below by adding the nonlinear terms to (8.37):

$$\begin{bmatrix} X_1(s) \\ X_2(s) \end{bmatrix} = \begin{bmatrix} G_{11}(s) & 0 \\ G_{21}(s) & G_{22}(s) \end{bmatrix} \begin{bmatrix} U_1(s) \\ U_2(s) \end{bmatrix} - \begin{bmatrix} f_{c1}(\dot{x}_1) + \Delta f_1(\dot{x}_1) \\ f_{c2}(\dot{x}_2) + \Delta f_2(\dot{x}_2) \end{bmatrix} \quad (8.38)$$

8.2.2.3 Parameter Estimation

8.2.2.3.1 Linear Term on Principal Diagonal

The transfer functions on the principal diagonal of the transfer function matrix shown in (8.38) describe the primary characteristic and dynamics of each USM stage, respectively. With the help of the System Identification Toolbox of MATLAB, the parameters are estimated as $a_1 = 308.5$, $b_1 = 2768$, $c_1 = 4272$, $a_2 = 226$, $b_2 = 3312$, and $c_2 = 2516$. Hence, the primary linear terms of the 2-DOF stage are given by

$$G_{11}(s) = \frac{4272}{s^2 + 308.5s + 2768} \quad (8.39)$$

$$G_{22}(s) = \frac{2516}{s^2 + 226s + 3312} \quad (8.40)$$

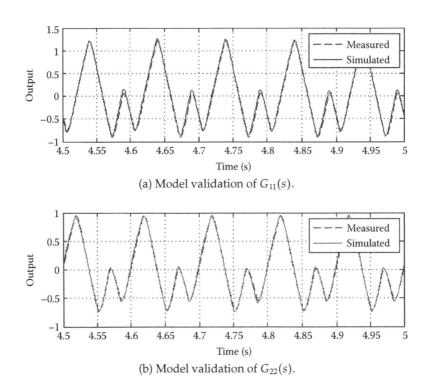

(a) Model validation of $G_{11}(s)$.

(b) Model validation of $G_{22}(s)$.

FIGURE 8.23
Model validation.

The model validation is shown in Figure 8.23. As can be seen, the simulated outputs of both models tally well with the actual measured outputs. The differences are more evident at the local maximum and minimum points, but they are very small and acceptable. Thus, the computed linear model is acceptable and can be used in the controller design.

8.2.2.3.2 Coupling Component

The coupling component of this 2-DOF system is described by the transfer function $G_{21}(s)$. From Figure 8.21, it behaves like a slow "drift." A second-order system is used to approximate the coupling phenomenon. With the help of the System Identification Toolbox of MATLAB, the parameters are estimated as $a_3 = 3804$, $b_3 = 90.06$, and $c_3 = 3.676$. Hence, $G_{21}(s)$ can be written as

$$G_{21}(s) = \frac{3.676}{s^2 + 3804s + 90.06} \tag{8.41}$$

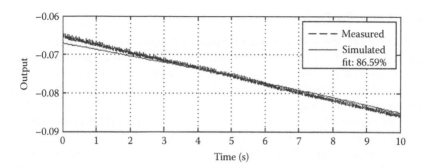

FIGURE 8.24
Model validation of $G_{21}(s)$.

The comparison of the actual measured output and the simulated output is shown in Figure 8.24. As can be seen, the slope of the simulated steady-state output is almost the same as the changing rate of the coupling effect. Therefore, the derived transfer function G_{21} is a good approximation of the coupling effect.

8.2.2.3.3 Nonlinear Term

To identify the nonlinear Coulomb friction model of each USM stage, a slow triangular wave with a frequency of 0.5 Hz and amplitude of ±4V is applied as analog input to the open loop system. From the experiments, it is found that the friction f_c is not symmetric along the forward and backward directions. Thus, the Coulomb friction can be written as

$$f_c(\dot{x}) = bu_c(\dot{x}) = [\sigma\,\text{sgn}(\dot{x}) - \delta]d = \begin{cases} (\sigma - \delta)d, & \dot{x} > 0 \\ 0, & \dot{x} = 0 \\ (-\sigma - \delta)d, & \dot{x} < 0 \end{cases} \quad (8.42)$$

where σ is a coefficient, sgn is a sign function, and δ and d are constants. Moreover, the constant d is related to the payload and the mover's weight.

During the experiments, the Z-axis stage begins to move when the input is around –2.993 V or 2.316 V, and the X-axis stage begins to move when the input is around –3.175 V or 3.115 V. The constants d are $d_1 = 3844.8$ and $d_2 = 2516$, respectively. Therefore, the estimated constants are $\sigma_1 = 2.6545$, $\delta_1 = 0.3385$, $\sigma_2 = 3.145$, and $\delta_2 = 0.03$. The nonsymmetric friction term of each stage can be described by

$$f_{c1}(\dot{x}_1) = 10206.02\text{sgn}(\dot{x}_1) - 1301.46 \quad (8.43)$$

$$f_{c2}(\dot{x}_2) = 7912.82\text{sgn}(\dot{x}_2) - 75.48 \quad (8.44)$$

8.2.3 Controller Design

In this chapter, the widely used PID feedback controller is chosen to be the main controller for the linear term since the PID controller involves simple operations that are easy to understand. The nonlinear term is compensated by a sign function and the sliding mode control. Moreover, because we want to get a fast system response and require the overshoot of the response to be as small as possible, an optimal control design method (LQR) can assist the PID tuning to obtain local optimal PID parameters.

In this section, an LQR-assisted PID control law will be presented at first, followed by a presentation of the nonlinear compensation method. Finally, the control of a 2-DOF USM stage will be illustrated.

8.2.3.1 LQR-Assisted PID Control

The following PID control law is used for the feedback control:

$$u(t) = K_p(x_d(t) - x(t)) + K_i \int_0^t e_1(\tau)d\tau + K_d \frac{d(x_d(t) - x(t))}{dt} \quad (8.45)$$

To apply the optimal control directly, the system model needs to be changed to an error model. Thus, define the position error $e_1(t) = x_d(t) - x(t)$. So the integral error is given by $e_0(t) = \int_0^t e_1(\tau)d\tau$ and the derivative error (velocity error) is given by $e_2(t) = \dot{e}_1(t)$. Since the linear term of the USM stage is a second-order system, choose the errors as the states, i.e., $E(t) = [e_0(t), e_1(t), e_2(t)]^T$, and we have

$$\dot{E}(t) = AE(t) - Bu(t) + B\left[\frac{f(\dot{x}) + \ddot{x}_d(t) + a\dot{x}_d(t) + bx_d(t)}{c}\right] \quad (8.46)$$

with

$$A = \begin{bmatrix} 0 & 1 & 0 \\ 0 & 0 & 1 \\ 0 & -a & -b \end{bmatrix}, B = \begin{bmatrix} 0 \\ 0 \\ c \end{bmatrix}$$

From this model, the dominant linear part is

$$\dot{E}(t) = AE(t) - Bu(t) \quad (8.47)$$

which is controllable due to the nonzero c.

As can be seen in (8.47), it can be observed that the vector X contains the PID term, which should be determined for the application. The PID controller is then converted to an equivalent state feedback controller. In this chapter, the PID parameters are obtained by using the LQR technique. Generally, the optimal LQR control is based on the following index:

$$J = \int_0^t [E(\tau)^T QE(\tau) + ru(\tau)^T u(\tau)]d\tau \quad (8.48)$$

where Q is the weighting matrix, normally chosen as a diagonal matrix, i.e., $Q = diag\{q_1, q_2, q_3\}$.

Consider the PID control structure and (8.47); the state feedback control form is taken as

$$u_l(t) = KE(t) \qquad (8.49)$$

where K is the feedback gain, $K = -r^{-1}B^T P = [k_1, k_2, k_3]^T$, and $P = \begin{bmatrix} p_{11} & p_{12} & p_{13} \\ p_{21} & p_{22} & p_{23} \\ p_{31} & p_{32} & p_{33} \end{bmatrix} > 0$ is the solution of the Riccati equation, shown below:

$$A^T P + PA - PBr^{-1}B^T P + Q = 0 \qquad (8.50)$$

Thus, the feedback controller is

$$u_l(t) = k_1 e_0(t) + k_2 e_1(t) + k_3 e_3(t)$$
$$= -r^{-1}cp_{31}e_0(t) - r^{-1}cp_{32}e_1(t) - r^{-1}cp_{33}e_2(t) \qquad (8.51)$$

Actually, the feedback gain K contains the PID parameters, i.e., the proportional gain $K_p = k_2$, the integral gain $K_i = k_1$, and the derivative gain $K_d = k_3$.

8.2.3.2 Nonlinear Compensation

The control law in (8.51) does not consider the effects of the nonlinear term. Referring to (8.30), (8.42), and (8.46), the nonlinear term is

$$\dot{E}(t) = B \left[\frac{\sigma\, \text{sgn}(\dot{x})d - \delta d + \Delta f(\dot{x})}{c} \right] \qquad (8.52)$$

The equation (8.52) can also be decomposed into two parts: the structured component $B[(-\sigma\,\text{sgn}(\dot{x})d + \delta d)/c]$ and the uncertain component $B[\Delta f(\dot{x})/c]$.

For the structured component, it is effectively eliminated with a sign function as shown below:

$$u_{f_c}(t) = \frac{\sigma\, \text{sgn}(\dot{x})d - \delta d}{c} \qquad (8.53)$$

For the uncertain component, a sliding mode control law (8.54) is proposed to reject it.

$$u_s(t) = \hat{k}_s \text{sgn}(E(t)^T PB) \qquad (8.54)$$

where $\hat{k}_s$ is used to adaptively estimate the amplitude of the uncertain term in the system, and that is given by

$$\dot{\hat{k}}_s = \rho_1 |E^T PB| - \rho_0 \hat{k}_s \qquad (8.55)$$

where the parameters ρ_0 and ρ_1 are adaptive gains.

Combining (8.53) and (8.54), the nonlinear compensation is

$$u_{nl}(t) = u_{f_c}(t) + u_s(t)$$
$$= \frac{\sigma \operatorname{sgn}(\dot{x})d - \delta d}{c} + \hat{k}_s \operatorname{sgn}(E(t)^T P B) \tag{8.56}$$

8.2.3.3 Control of 2-DOF USM Stage

In this subsection, the PID controllers with nonlinear compensation for both stages are designed according to the design method as presented in previous subsections. Furthermore, the controller design for the Z-axis USM stage can completely follow the design method since it is only affected by the input u_1. However, for the X-axis USM stage, both inputs u_1 and u_2 have effects on its output, which means it is coupled. Therefore, a model modification based on the coupled linear model for decoupling is introduced.

8.2.3.3.1 Controller Design for Z-Axis Stage

First, we design the controller for the Z-axis. Considering nonlinear effects in the system, the model of the Z-axis is given by

$$\ddot{x}_1(t) = -b_1 x_1(t) - a_1 \dot{x}_1(t) + c_1 u(t) + f_1(\dot{x}_1) \tag{8.57}$$

Define the error $e_{z1}(t) = x_{1d}(t) - x_1(t)$. The integral error is given by $e_{z0}(t) = \int_0^t e_{z1}(\tau)d\tau$ and the derivative error is given by $e_{z2}(t) = \dot{e}_{z1}(t)$. Let $E_1(t) = [e_{z0}(t), e_{z1}(t), e_{z2}(t)]^T$, and we have

$$\dot{E}_1(t) = A_1 E_1(t) - B_1 u_1(t)$$
$$+ B_1 \left[\frac{f_1(\dot{x}_1) + \ddot{x}_{1d}(t) + a_1 \dot{x}_{1d}(t) + b_1 x_{1d}(t)}{c_1} \right] \tag{8.58}$$

where

$$A_1 = \begin{bmatrix} 0 & 1 & 0 \\ 0 & 0 & 1 \\ 0 & -b_1 & -a_1 \end{bmatrix}, B_1 = \begin{bmatrix} 0 \\ 0 \\ c_1 \end{bmatrix}$$

From this model, the dominant linear part is

$$\dot{E}_1(t) = A_1 E_1(t) - B_1 u_1(t) \tag{8.59}$$

Apply the LQR-assisted PID controller design method to (8.59), and we can obtain the controller for the linear term as shown below:

$$u_{1l}(t) = K_1 E_1(t) = r_1^{-1} B_1^T P_1 E_1(t) \tag{8.60}$$

For the nonlinear term, apply (8.56) to the system, and we have

$$u_{1nl}(t) = \frac{\sigma_1 \text{sgn}(\dot{x}_1)d_1 - \delta_1 d_1}{c_1} + \hat{k}_{1s}\text{sgn}(E_1(t)^T P_1 B_1) \tag{8.61}$$

with $\dot{\hat{k}}_{1s} = \rho_{11}|E_1^T P_1 B_1| - \rho_{10}\hat{k}_{1s}$.

In summary, the control law for the Z-axis stage is given by

$$u_1(t) = r_1^{-1}B_1^T P_1 E_1(t) + \frac{[\sigma_1\text{sgn}(\dot{x}_1(t)) - \delta_1]d_1}{c_1} + \hat{k}_{1s}\text{sgn}(E_1(t)^T P_1 B_1) \tag{8.62}$$

8.2.3.3.2 Controller Design for X-Axis USM Stage

Next, let us see the X-axis control. From (8.36), we have

$$X_2(s) = \frac{c_2}{s^2 + a_2 s + b_2}U_2(s) + \frac{c_3(s^2 + a_2 s + b_2)}{(s^2 + a_2 s + b_2)(s^2 + a_3 s + b_3)}U_1(s)$$

$$= \frac{c_2}{s^2 + a_2 s + b_2}U_2(s)$$

$$+ \frac{c_3(s^2 + a_3 s + b_3) + c_3(a_2 - a_3)s + c_3(b_2 - b_3)}{(s^2 + a_2 s + b_2)(s^2 + a_3 s + b_3)}U_1(s)$$

$$= \frac{c_2 U_2(s) + c_3 U_1(s) + c_3 \tilde{U}_1(s)}{s^2 + a_2 s + b_2} \tag{8.63}$$

where $\tilde{U}_1(s) = \frac{(a_2 - a_3)s + (b_2 - b_3)}{s^2 + a_3 s + b_3}U_1(s) = F(s)U_1(s)$ acts like a filter. Considering nonlinear effects in the X-axis, we have the following model:

$$\ddot{x}_2(t) = b_2 x_2(t) + a_2 \dot{x}_2(t) + c_2 u_2(t) + c_3 u_1(t) + c_3 \tilde{u}_1(t) + f_2(\dot{x}_2) \tag{8.64}$$

Let $E_2 = [e_{x0}, e_{x1}, e_{x2}]^T$. The above equation becomes

$$\dot{E}_2(t) = A_2 E_2(t) - B_2 u_2(t) + B_2 \left[\frac{f_2(\dot{x}_2) + \ddot{x}_{2d}(t) + a_2 \dot{x}_{2d}(t) + b_2 x_{2d}(t)}{c_2} \right]$$

$$+ B_2 \left[\frac{c_3 u_1(t) + c_3 \tilde{u}_1(t)}{c_2} \right] \tag{8.65}$$

where

$$A_2 = \begin{bmatrix} 0 & 1 & 0 \\ 0 & 0 & 1 \\ 0 & -b_2 & -a_2 \end{bmatrix}, B_2 = \begin{bmatrix} 0 \\ 0 \\ c_2 \end{bmatrix}$$

From the model, the coupled linear part is

$$B_2 \left[\frac{c_3 u_1(t) + c_3 \tilde{u}_1(t)}{c_2} \right] \tag{8.66}$$

This part can be eliminated by the following equation:

$$u_{2de}(t) = \frac{c_3[u_1(t) + \tilde{u}_1(t)]}{c_2} \tag{8.67}$$

The proposed controller is given by

$$u_2(t) = u_{2de}(t) + u_{2d}(t) \tag{8.68}$$

where $u_{2d}(t)$ is the control law that needs to be designed.

Substitute the proposed controller into (8.65). It is observed that the system (8.65) becomes the following form:

$$\dot{E}_2(t) = A_2 E(t) - B_2 u_{2d}(t) + B_2 \left[\frac{f_2(\dot{x}_2) + \ddot{x}_{2d} + a_2 \dot{x}_{2d} + b_2 x_{2d}}{c_2} \right] \tag{8.69}$$

The new system (8.69) is only affected by $u_{2d}(t)$, which means it is decoupled. Moreover, it should be noted that this form is the same as in (8.58). Therefore, the controller design method presented in Section 8.2.3.1 can be used for this decoupled system. Thus, we have a similar control law $u_{2d}(t)$ given by

$$u_{2d}(t) = K_2 E_2(t) + \frac{\sigma_2 \text{sgn}(\dot{x}_2) d_2 - \delta_2 d_2}{c_2} + \hat{k}_{2s} \text{sgn}(E_2(t)^T P_2 B_2) \tag{8.70}$$

with $K_2 = r_2^{-1} B_2^T P_2$ and $\dot{\hat{k}}_{2s} = \rho_{21} |E_2^T P_2 B_2| - \rho_{20} \hat{k}_{2s}$.

In summary, the control law for the X-axis stage is given by

$$u_2(t) = \frac{c_3 u_1 + c_3 \tilde{u}_1}{c_2} + r_2^{-1} B_2^T P_2 E_2(t)$$

$$+ \frac{\sigma_2 \text{sgn}(\dot{x}_2) d_2 - \delta_2 d_2}{c_2} + \hat{k}_{2s} \text{sgn}(E_2(t)^T P_2 B_2) \tag{8.71}$$

8.2.3.3.3 Overall Controller

Combine (8.62) and (8.71). The overall controller for the 2-DOF USM stage is shown in Figure 8.25.

8.2.4 Experimental Results

The experimental setup is shown in Figure 8.26. The system consists of a 2-DOF USM stage, two motor drives, and a PC with a dSPACE 1104 control card. The sampling time is 0.0005 s.

The PID controller is designed based on the LQR approach. The weighting matrices for both axes are chosen as $Q_1 = Q_2 = Q = \text{diag}\{10^4, 10^3, 10^{-5}\}$ and $r_1 = r_2 = r = 1$. Therefore, the PID parameters for both axes can be calculated by (8.50), giving $K_1 = [100, 31.4248, 0.0690]$ and $K_2 = [100, 30.8996, 0.0908]$.

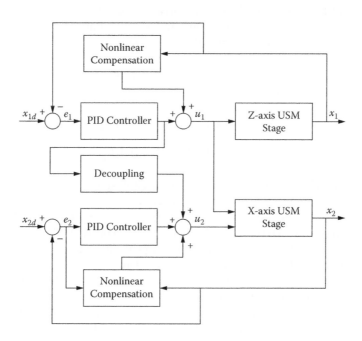

FIGURE 8.25
Block diagram of the overall controller.

FIGURE 8.26
System setup.

The eigenvalues of the closed-loop system are $\{-300.02 \pm 212.38i, -3.16\}$ and $\{-225.70 \pm 169.38i, -3.16\}$, which contain real parts that are all negative, i.e., $Re(\lambda\{A - BK\} < 0)$. This implies that the designed PID control system is stable. On the other hand, the adaptive gains are chosen as $\rho_{10} = \rho_{20} = 0.01$ and $\rho_{11} = \rho_{21} = 0.0001$; then the adaptive control part for nonlinear compensation is given by

$$\dot{k}_{1s} = 0.0001|X_1^T P_1 B_1| - 0.01\hat{k}_{1s} \tag{8.72}$$

$$\dot{k}_{2s} = 0.0001|X_2^T P_2 B_2| - 0.01\hat{k}_{2s} \tag{8.73}$$

In order to achieve the cycle path by the 2-DOF stage, the reference signal to the Z-axis stage is chosen as a sine wave, while a cosine wave reference signal is chosen for the X-axis stage. The amplitude and frequency of the input signal are the same, ± 0.6 mm and 15 Hz. The reason why 15 Hz is used is because fast motion is required; i.e., the surgical time should be as short as possible so as to reduce the damage to the eardrum.

Without decoupling, the control performance of one cycle by using the LQR-assisted PID controller without and with nonlinear compensation is shown in Figure 8.27. As can be seen, there is only a small portion of the path that can be tracked by using the pure PID controller, while the path is tracked well by using the PID controller with compensation.

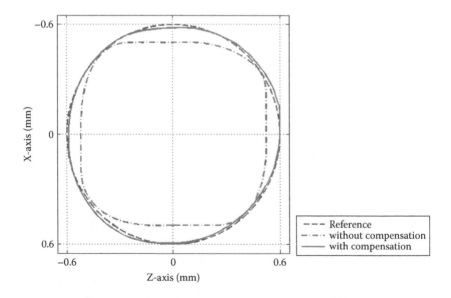

FIGURE 8.27
Control performance of one cycle by using the PID controller without and with compensation without decoupling.

TABLE 8.2

Tracking Errors by Using Different Control Laws

Control Law	Pure PID	With Compensation without Decoupling	With Compensation and Decoupling
$\bar{e}$	0.1575	0.0550	0.0536

In order to verify the effects of the nonlinear compensation and the decoupling method, use the two-norm $\| \bullet \|_2$ of the errors in both axes to measure the 2-DOF errors. The computational formula of two-norm is defined by

$$\| e \|_2 = \sqrt{e_{z1}^2 + e_{x1}^2} \tag{8.74}$$

and the mean of $\| e \|_2$ is denoted by $\bar{e}$ and used to evaluate the tracking performance.

$$\bar{e} = \frac{1}{n} \sum_{i=1}^{n} (\| e \|_2) \tag{8.75}$$

In another experiment, the reference signal is kept unchanged, running 300 cycles (20 s) by using different control laws: (1) PID controller without nonlinear compensation and decoupling (pure PID), (2) PID controller with compensation but without decoupling, and (3) PID controller with compensation and decoupling. The control performance is shown in Table 8.2. As can be seen, the error of the controller with nonlinear compensation is much smaller than the pure PID case, which is one-third of the error of the controller without compensation and less than 10% of the amplitude of the reference signal. The decoupling controller helps to reduce the error by about 2.5%. Furthermore, in some cases, the decoupling controller can reduce the error by 5–10% when the coupling effect is stronger, e.g., the motors are not mounted properly. It can be concluded that the nonlinear compensation can greatly improve the performance, and the error can be further reduced upon adding the decoupling controller.

Finally, the proposed PID controller with decoupling and nonlinear compensation is applied for grommet insertion on a mock membrane. The grommet is successfully inserted on the membrane. The desired path, the system output, and the mock membrane before and after the grommet is inserted are shown in Figure 8.28. The maximum and the mean of the two-norm of the error are 0.6544 and 0.0248 mm, respectively. The overall insertion time is 0.6585 s (less than 1 s). Besides that, the deformation caused using the designed path is much smaller than directly inserting, so that the damage to the membrane is minimal.

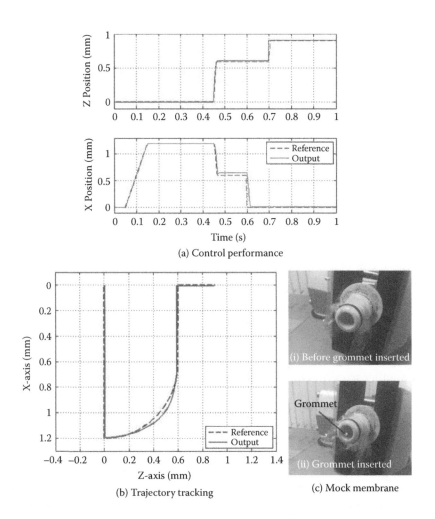

FIGURE 8.28
Control performance of grommet insertion.

8.3 Vision-Based Tracking and Thermal Monitoring of Nonstationary Targets

Computer imaging technology has grown rapidly in recent decades and pervaded every aspect of our daily lives. This trend generates ample opportunities for the development of new image and vision-based applications. It has redefined the concept of surveillance and design of human–machine interaction. This section aims to develop image and vision systems for CNC machine leveraging on precise actuators to track a nonstationary moving tool with vision on the feedback. A decoupled tracking and thermal monitoring system is developed that can be used on nonstationary targets of closed systems such as machine tools.

8.3.1 Computer Imaging Technologies

Computer imaging is a fascinating and exciting research area nowadays. Visual information, transmitted in the form of digital images, is becoming an important means of research. Computer imaging can be defined as the acquisition and processing of visual information by computer, which can be divided into two primary categories:

- Image processing
- Computer vision

These two categories are not totally separate and distinct [24]. There are no clear boundaries in the continuum from image processing at the one end to computer vision at the other.

Image processing: Image processing is a form of computer imaging where the application involves a human being in the visual loop [25]. In other words, the images are to be examined and acted upon by people. Major application fields of image processing include medical imaging [26], industrial inspection, and astronomical observation. With the rapid development of computer and image technology and the increasing demand for picture and image technology, image processing has entered the motion control field and improved the control quality, so that the level of the vision-guided robotic system has greatly improved by using the image operation and analysis. Other ongoing research areas include movement extraction and recognition from images. Application fields include motion positioning extraction from the industrial environment, natural scene images, and videos. A powerful image processing system for vehicle navigation has been developed based on natural scene image recognition [27]. Automatic video caption translation software can be designed using caption extraction and recognition schemes for every frame of a video [28].

Computer vision: Computer vision is the other form of computer imaging where the application does not involve a human being in the visual loop. In other words, the images are examined and acted upon by a computer. Although people are involved in the development of the system, the final application requires a computer to use the visual information directly. One of the major topics within the field of computer vision is image analysis.

The field of computer vision may be best understood by considering different types of applications. Many of these applications involve tasks that either are tedious for people to perform, require work in a hostile environment, require a high rate of processing, or require access and use of a large database of information. Computer vision systems are used in many and various types of environments—from manufacturing plants to hospital surgical suites to the surface of Mars. The most important task of computer vision systems is automated visual inspection (AVI) [29], which can be used for the purpose of measurements, gauging, integrity checking, and quality control. In the field of measurements, the gauging of small gaps [30], measurement of object dimensions, alignment of the components, and analysis of crack formation are common applications. For example, the computer vision system will scan manufactured items for defects and provide control signals to a robotic manipulator to remove defective parts automatically [31]. During the automotive assembly, a vision-guided robot identifies and sorts the different parts. Computer vision systems are also used in many different areas within the medical and pharmacological community, with the only certainty being that the types of applications will continue to grow. Current examples of medical systems being developed include systems to diagnose skin tumors automatically [32], systems to aid neurosurgeons during brain surgery, systems to perform clinical tests, and systems for automatic cell injection. Computer vision systems that are being used in the surgical suites have already been used to improve surgeons' ability to see what is happening in the body during surgery, and consequently improve the quality of medical care available [33]. Computer vision systems are also used to manipulate whole transportation systems in an automatic and intelligent way.

Another term that has a similar meaning as computer vision is machine vision [34]. Machine vision is concerned with the engineering of integrated mechanical-optical-electronic-software systems for examining natural objects and materials. Although it uses similar computational techniques, it does not necessarily involve a device that is regarded as a computer.

8.3.2 Decoupled Tracking and Thermal Monitoring

In this section, a decoupled tracking and thermal monitoring system using computer imaging technology is presented that can be used on nonstationary targets of closed systems such as machine tools.

The typical machine surveillance scheme was done by using sensor monitoring, data analyzing, and modeling. However, the sensor set has obvious shortcomings, such as low sampling rate or power constraint. A thermal camera was also used in such applications. Apart from its high price, most thermal cameras can only save the highest temperature values inside the testing scope unless other software is provided. In CNC machine surveillance, the highest temperature is usually generated by metal debris, which is not the point of interest, instead of the milling tool. Manual work is needed to find the temperature of the milling tool through a lookup table provided by the manufacturer. In view of this fact, a thermal camera is not the optimized solution since our objective is to real-time monitor the temperature of the working tool. The following subsections will present a vision-based tracking algorithm such that the thermometer carried by the linear motor controller can follow the moving tool and read its temperature value online, which overcomes the above-mentioned limitations.

8.3.2.1 Overall System Configuration

The monitored system (see Figure 8.29) is a milling machine, and the condition of the cutting tool will be monitored by inferring the temperature of the cutting point. The temperature measurements are not naturally available, and it is not possible to mount temperature sensors on the tool or the workpiece, so a noncontact mode of temperature measurement will be necessary. An infrared thermometer with a built-in laser will be used for this purpose. In addition, the tool is nonstationary throughout the cutting process. Hence, the thermometer must be moved in tandem with the milling tool to track the tool. The thermometer is thus mounted on a decoupled motion stage. As the milling machine is a closed system, its encoder signals are not readily available for use by the motion stage carrying the thermometer to track the tool. A video camera is used instead to record the motion of the milling tool,

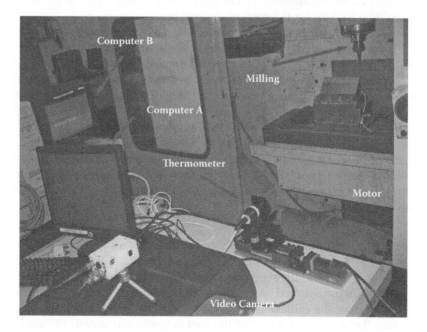

FIGURE 8.29
Overall system configuration.

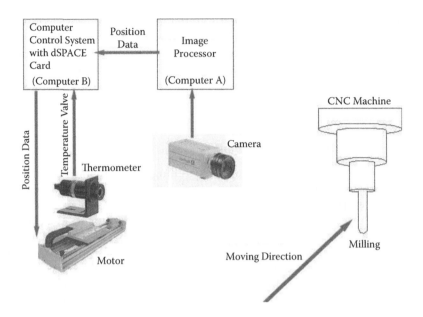

FIGURE 8.30
Vision-assisted servo system.

and image processing algorithms will extract the position information related to the milling tool tip for the motion stage to track.

There are thus three major subsystems in the overall system. The vision and image processing system will process real-time images to extract the position of the milling tool. The noncontact infrared thermometer will efficiently provide temperature measurements of a selected spot on the milling tool. The computer control system will move the thermometer in tandem with the movement of the milling tool based on the position feedback from the vision system. Figure 8.30 shows the interaction among these three subsystems.

The development of the monitoring system is based on two assumptions. First, the camera is stationary. Second, the illumination condition is consistent and will not experience considerable changes. The initial setup will mainly involve the calibration of the camera and proper mounting of the thermometer on the motion stage.

> **Camera calibration:** The camera should be able to resolve the actual motion of the milling tool from the images captured. Therefore, proper calibration should first be done as follows. The camera (screen size: 768×576) is mounted on a table located at a fixed distance of 500 mm away from the milling tool for safety reasons, and it is adjusted so that the maximum travel of the tool is contained within the view of the camera. Within this span, the tool is manually displaced by

FIGURE 8.31
Mounting of the infrared thermometer.

a fixed distance D_{CR}, and the equivalent displacement of the image captured by the camera is denoted as D_{CV}. The transformation coefficient k can be computed from $D_{CR} = k \times D_{CV}$. With this calibration, subsequently, the displacement of the tool D_R can be calculated based on the image displacement D_V through

$$D_R = \frac{D_{CR}}{D_{CV}} D_V \tag{8.76}$$

Thermometer mounting: The thermometer used is an OS550A industrial infrared thermometer from Omega Engineering. It has a distance-to-target ratio of 609:8.9 mm (diameter of the tip of the cutting tool is 12 mm). Thus, the thermometer has to be adequately fixed to yield accurate temperature measurements. First, the linear motor (on which to mount the thermometer) is located at a position 609 mm away from the milling tool and aligned with the direction of travel of the tool. The thermometer is mounted on the motor such that its laser is directed at the spot of the tool to measure its temperature (Figure 8.31).

In the ensuing sections, the three main subsystems of the overall monitoring system will be elaborated in detail.

8.3.2.2 Vision and Image Processing System

The purpose of the vision and image processing system is to detect and track the milling tool so that the thermometer can be driven by the linear motor to follow the milling tool and continuously monitor its temperature. The milling tool will be detected through the first few frames and subsequently tracked by updating its position iteratively. Moving object extraction is an active field of computer vision, and it has wide practical applications in industrial monitoring systems [24]. Effectively detecting and tracking a target object from video sequences is the main task in the proposed surveillance system. Currently three classical algorithms can be used in video surveillance systems. In-depth reviews on these methods can be found in [35]. Optical flow [36] is the most computationally intensive method, which is not suitable for real-time processing in many practical applications. Background subtraction [26], which is usually used in traffic monitoring, detects moving objects by comparing the current frame with a reference. Typically, the object to be tracked only occupies a small portion of the frame. However, for the target application in the section, this method is not applicable since the background reference is not available, and to obtain one will entail moving away the machine, which is cumbersome. Frame difference [37] is the simplest method and easy to implement. It creates a binary image by comparing two consecutive frames; if the difference of a pixel is higher than a threshold, then the pixel will result in 1 and 0 otherwise. The main drawback of frame difference is its sensitivity to Gaussian noise. Considering the time-critical requirement of our monitoring system, the tool detection algorithm will be based on the frame difference method. The process flowchart is shown in Figure 8.32.

The basic idea of the proposed algorithm is to highlight the salient moving gradient [35] between frame sequences, which can be expressed by

$$\frac{dI(x,y,t)}{dt} = \frac{\partial I(x,y,t)}{\partial x}\frac{\partial x}{\partial t} + \frac{\partial I(x,y,t)}{\partial y}\frac{\partial y}{\partial t} + \frac{\partial I(x,y,t)}{\partial t}$$

$$= G_x\frac{\partial x}{\partial t} + G_y\frac{\partial y}{\partial t} + I_t \tag{8.77}$$

where $G_x\frac{\partial x}{\partial t}$ and $G_y\frac{\partial y}{\partial t}$ represent the moving gradients along the x and y directions, respectively. The charge-coupled device (CCD) camera captures 30 frames per second. After acquisition, the edge images $E(x,y,t-\Delta)$ of $I(x,y,t-\Delta)$ are created using the same kernel as reported in [38]. They are computed as follows:

$$E(x,y,t) = \begin{cases} 255 & if\ I(x,y,t)\ is\ not\ edge\ pixel \\ 0 & if\ I(x,y,t)\ is\ edge\ pixel \end{cases} \tag{8.78}$$

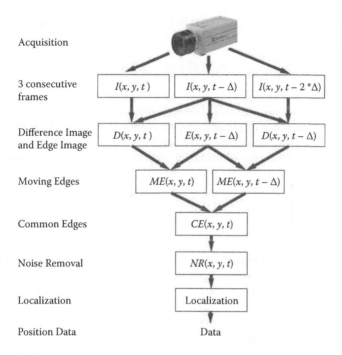

FIGURE 8.32
Process flowchart.

Meanwhile, two difference images, $D(x, y, t)$ and $D(x, y, t - \Delta)$, are generated by comparing the illumination difference $\Delta L(x, y, t)$, computed as follows:

$$L(x, y, t) = 0.299 \times I(x, y, t).R + 0.587 \times I(x, y, t).G + 0.114 \times I(x, y, t).B$$

$$D(x, y, t) = \begin{cases} 0 & if \ |L(x, y, t) - L(x, y, t - \Delta)| > T \\ 1 & if \ |L(x, y, t) - L(x, y, t - \Delta)| < T \end{cases}$$

A proper value of threshold T should be set based on illumination and contrast. In this application, a suitable T value is found to be 5. Moving edges $ME(x, y, t)$ can be computed as

$$ME(x, y, t) = D(x, y, t) \times E(x, y, t) \tag{8.79}$$

As discussed before, the frame difference method is sensitive to noise, and a simple moving edge operation actually has the sum of edges of both frames, which will contain the residual of previous contours. Thus, common edge operation is added to eliminate this residual. Common edge $CE(x, y, t)$ is regarded as an AND operation of two consecutive moving edges, which can be computed as

$$CE(x, y, t) = \frac{ME(x, y, t) \times ME(x, y, t - \Delta)}{255} \tag{8.80}$$

$CE(x, y, t)$ is a binary image with black moving edges and a white motionless part. Noise or tiny moving objects are also highlighted in $CE(x, y, t)$. Therefore, a noise removal step is needed to eliminate them. A statistical approach is used to realize this noise elimination, through the equations below:

$$NR(x, y, t) = \begin{cases} 255 & if \ NR(x, y, t) = 0 \\ & and \ \sum_{n=-3}^{n=3} \sum_{m=-3}^{m=3} \frac{NR(x+n, y+m, t)}{255} > 3 \\ NR(x, y, t) & otherwise \end{cases} \tag{8.81}$$

An isolated black pixel is regarded as isolated noise, and it will turn white if it has more than three white neighboring pixels.

Following the above steps as summarized in Figures 8.33–8.39, the contour of the milling tool will be obtained. A feature-based localization method is then used to locate the position of the milling tool. In the proposed algorithm, the edges of the milling tool can be located from two $2 \times L$ rectangles, where L is the estimated length of the milling tool. The final position of the tool can be obtained from the midpoint between the two sides. A template-matching algorithm will be required to locate an irregularly shaped milling

FIGURE 8.33
Moving object extraction: original image.

FIGURE 8.34
Moving object extraction: difference image $D(x, y, t)$.

FIGURE 8.35
Moving object extraction: difference image $D(x, y, t - \Delta)$.

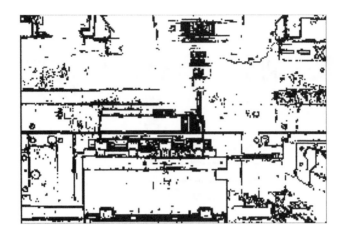

FIGURE 8.36
Moving object extraction: edge image.

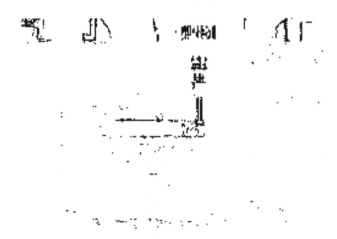

FIGURE 8.37
Moving object extraction: common edge.

FIGURE 8.38
Moving object extraction: noise removal result.

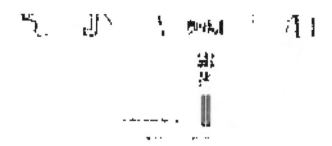

FIGURE 8.39
Moving object extraction: localization.

tool. The salient gradient can be found horizontally within the estimated vertical range of the milling $[Y_1, Y_2]$, which can be expressed by the following pseudocode:

```
Begin:
x = 0, Max1 = 0, Max2 = 0, Side1 = 0, Side2 = 0;
Loop: Count the number of edge pixels PiCt within rectangle
[(x,Y₁), (x +1,Y₂)]
if (PiCt>Max1) {Max1 = PiCt;Side1 = x;}
elseif (PiCt > Max2){Max2 = PiCt;Side2 = x;}
MillPosition = Side1+Side2
               2
x++;
if (x = Width of the screen-1) End;
else go to Loop;
```

The milling tool can be located based on the above-described algorithm through the first few frames. To reduce the amount of computation necessary, the search to [MillPosition – 5, MillPosition + 5] can be narrowed along the horizontal direction in the following frames.

8.3.2.3 Noncontact Temperature Measurement System

Conventional temperature sensors such as thermistors and thermocouples are not applicable to measure temperature on parts that are moving or are in contact with other parts. The target application, addressed in this section, is one such example. A noncontact temperature measurement system will be necessary in these cases. Thermal cameras (Figure 8.40(a)) can fill these gaps, and they are finding their ways to a variety of applications. However, one main drawback of using a thermal camera is the high cost associated with it. An alternate way to measure the temperature in a noncontact manner is to use an infrared thermometer with a built-in laser tracer (Figure 8.40(b)) to pinpoint the spot on the part to be measured. However, such a thermometer can only detect temperature of a spot instead of an extended area. The difficulty becomes even more pronounced when the part is nonstationary. The challenge

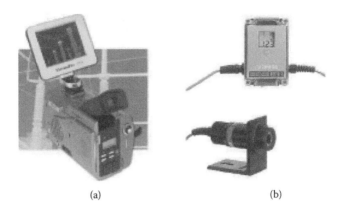

(a)   (b)

FIGURE 8.40
Thermal devices. (a) Thermal camera. (b) Infrared thermometer.

is thus to formulate a moving platform on which the thermometer is mounted to move in tandem with the moving part closely so as to obtain continuous and reliable temperature measurements from it.

8.3.2.4 Tracking Control of Linear Motor

In the target application, the infrared thermometer is carried on a linear motor (see Figure 8.31). It can measure the temperature of a moving object as long as the motion of the object is tracked sufficiently by the motor. Since the cutting tool is a moving target here, the tracking control system has to allow the motor to follow the tool based on the image position from the camera. The 1-DOF motor system under investigation is a mass constrained to move in one dimension with friction and periodic forces present between the mass and the supporting surface. The equation for this model is described as follows:

$$M\ddot{x} + D\dot{x} + F_l = u \tag{8.82}$$

where x denotes position; M, D, and F_l denote the mechanical parameters inertia, viscosity constant, and load force, respectively; and u denotes the control torque.

The PID control design is based on the error $e = x_d - x$, where x_d is the objective signal given by the image position whose range belongs to [X_Position -5, X_Position $+5$], as shown in the last section. Since

$$\frac{d}{dt}\int_0^t e\,d\tau = e \tag{8.83}$$

the standard PID control structure is given by

$$u = K_p e + K_i \int_0^t e\,d\tau + K_d \dot{e} \tag{8.84}$$

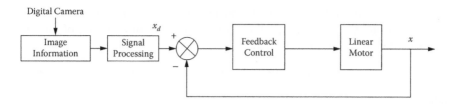

FIGURE 8.41
Control structure.

where K_p, K_i, and K_d are the PID control parameters. The trial-and-error method can be used to tune these parameters. Alternatively, advanced optimum control theory can be applied to tune PID control gains. For example, the PID feedback controller can be designed using the linear quadratic regulator (LQR) technique. The control structure is shown in Figure 8.41.

Since the image obtained from the camera includes time delay, the information cannot be used for the control loop directly. A signal processing technique is presented to solve this problem.

Consider the following model for the position information:

$$x_d(k) = x_d(k-1) + Tv(k-1) + w(k-1) \tag{8.85}$$

where $x_d(k)$ and $v(k)$ are actual moving position and velocity of the cutting tool, k is the sampling time of the control system, and $w(k)$ is the noise. Define $y(k)$ and $z(k)$ as the position and velocity obtained from the camera. From the previous section, it is known that

$$y(k) = x_d(k-h) \tag{8.86}$$

$$z(k) = v(k-h) \tag{8.87}$$

where h is the time delay due to the camera handling. Due to the delay measurement, it is necessary to develop a model to predict the current position information for the purpose of the motion control. To achieve this objective, we establish the following model:

$$\hat{x}_d(k) = \hat{x}_d(k-1) + Tv(k-1) \tag{8.88}$$

Since $z(k)$ is a delay signal, the position information is predicted assuming that the speed is constant during time $[k-h, k]$.

$$\hat{x}_d(k) = \hat{x}_d(k-h) + hTv(k-h)$$
$$= z(k) + hTv(k-h) \tag{8.89}$$

For a conventional PID feedback control system, the closed-loop system can be guaranteed to be stable for a nominal system. A stability analysis will be provided below.

Let the system state variables be assigned as $x_1 = \int_0^t e\,d\tau$, $x_2 = e$, and $x_3 = \dot{e}$. Denote $X = [x_1, x_2, x_3]^T$; the above equation can then be put into the equivalent state space formulation.

$$\dot{x} = Ax + Bu + B(-M\ddot{x}_d - D\dot{x}_d - F_l) \tag{8.90}$$

where

$$A = \begin{bmatrix} 0 & 1 & 0 \\ 0 & 0 & 1 \\ 0 & 0 & -D/M \end{bmatrix}, \ B = \begin{bmatrix} 0 \\ 0 \\ -1/M \end{bmatrix} \tag{8.91}$$

For PID control $u = Kx$ with $K = [K_i, K_p, K_d]$, the closed-loop system is given by

$$\dot{x} = (A + BK)x + B(-M\ddot{x}_d - D\dot{x}_d - F_l) \tag{8.92}$$

The PID controller can be tuned to stabilize the system by appropriate design of the matrix $A + BK$. Based on the designed PID control, the stability of the overall system is analyzed as follows:

Consider the Lyapunov function $V = X^T P X$. Thus, it follows that

$$\dot{V} = X^T[(A + BK)^T P + P(A + BK)]X + 2X^T P B(-M\ddot{x}_d - D\dot{x}_d - F_l) \tag{8.93}$$

Since x_d is a bounded image position, it may be a constant or smooth function. Thus, $\dot{x}_d$ and $\ddot{x}_d$ are also bounded, i.e.,

$$|(-M\ddot{x}_d - D\dot{x}_d - F_l)| \le f_B \tag{8.94}$$

with constant f_B. Since the matrix $A + BK$ is stable, there exists a $Q > 0$ such that the Lyapunov equation

$$(A + BK)P + P(A + BK) = -Q. \tag{8.95}$$

Therefore, it follows that

$$\dot{V} \le -\lambda_{min}(Q)||X||^2 + 2f_B||X|| \tag{8.96}$$

which is guaranteed negative as long as $||X|| > \frac{2f_B}{\lambda_{min}(Q)}$. This implies that X is uniformly ultimately bounded. An arbitrary small error of $||X||$ may be achieved by selecting large PID gains. This implies that the tracking control will drive the actual position $x \to x_d$ closely. This also implies that the temperature sensor will follow and sense the moving cutting tool.

8.3.2.5 Practical Issues

There are two main challenges with respect to milling tool tracking for temperature measurement on the fly due to the different dynamics in the three main components of the system: the camera, the thermometer, and the servo system. First, the requirement on motion tracking is high. The spot size of the thermometer is the same as the diameter of the tip of the milling tool. If the thermometer misses the target, the temperature will be a combination of the temperature of milling and ambient temperature, which can be expressed as

$$T = \frac{\alpha}{\alpha + \beta} T_{milling} + \frac{\beta}{\alpha + \beta} T_{Am} \tag{8.97}$$

where α and β are the sizes of the spot on the milling tool. This temperature measurement will not be accurate when the tracking of the tool is not done accurately. Second, to add to the first problem, the frequency of position updating from the camera (15 times per second) is much slower than motor dynamics (1000 times per second), which may lead to a lag effect. These issues will impose a constraint on the velocity of the milling tool so that adequate tracking and thus temperature measurements can be obtained.

The infrared thermometer has a time constant of $T_{ther} = 0.1$ s. The tracking time constant of the servo system is $T_{mot} = 0.01$ s. The camera's response time can also be regarded as the image processing time, $T_{cam} = 0.08$ s. Therefore, the total response time is $T_{tol} = 0.19$ s.

First, the maximum speed permissible will be determined. The thermometer can measure the temperature of a spot of 9 mm in diameter. Referring to Figure 8.42, when the tip of the milling tool moves away from point O, the servo system transporting the thermometer will experience a time delay equal to $T_{mot} + T_{cam} + T_{ther}$ before the tracking movement is completed and a temperature measurement is obtained. Over this period, the milling tool should not

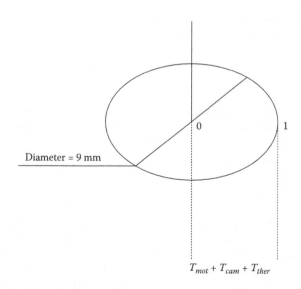

FIGURE 8.42
Maximum speed permissible.

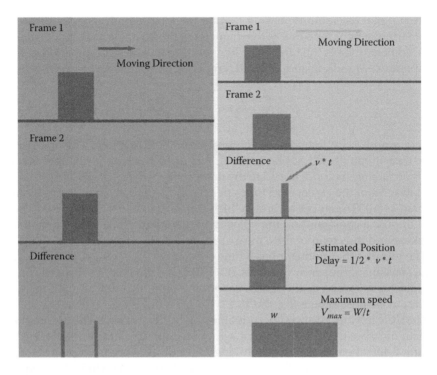

FIGURE 8.43
Calculation of minimum and maximum speeds.

move out of the laser spot (i.e., it should not move beyond point 1). Thus, the maximum speed of the milling tool can be obtained:

$$v'_{max} = \frac{4.5 \text{ mm}}{T_{mot} + T_{cam} + T_{ther}} = 23.7 \text{ mm/s} \qquad (8.98)$$

Second, from vision tracking point of view, since the method is based on frame difference, at least one pixel must be generated in the difference image $D(x, y, t)$, which is depicted in Figure 8.43. Suppose its moving speed is v and the time interval between two frames is Δt; the moving distance between two consecutive frames is thus $v \times \Delta t$. Based on $v \times \Delta t \geq 1$, minimum speed is $v_{min} = \frac{1}{\Delta t}$ pixel. Suppose the frequency is 10 Hz, the minimum speed is 10 pixels per second. The actual speed can be obtained using the equation in the camera calibration section. In the ball-milling machine experiment, the minimum speed of the tool is 1.5 mm/s. There is also a maximum speed for the image processing algorithm to function adequately. The maximum allowable speed is reached when the object in the second frame detaches from that in the first frame, so that max $v'_{max} = \frac{W}{\Delta t}$. W is the width of the object appearing on the screen. In the target application, this maximum speed of the tool is

27 mm/s. In the milling machine tested, a typical feed speed is 5 mm/s. The monitoring system is thus expected to be adequate to track the milling tool and yield temperature measurements.

8.3.2.6 Experimental Results

To illustrate the effectiveness of the proposed method, real-time experiments were conducted. The complete experiment was divided into three stages. In the first stage, a part-simulated experiment was conducted where we used a soldering iron (7 cm in diameter at the tip) to model a milling tool. The temperature of the soldering iron can be manually changed to simulate different milling situations. In the second stage, real experiments were conducted on ball-milling and face-milling machines, respectively. In the third stage, a thermal camera was used to test the accuracy of the above experimental data. In all the experiments, the infrared thermometer is fixed on a linear motor (Yaskawa SGML-01AF12) that can move at a maximum speed of 3000 rpm. The dSPACE control development and rapid prototyping system, in particular, the DS1102 board, is used. dSPACE integrates the whole development cycle seamlessly into a single environment. MATLAB/Simulink can be directly used in the development of the dSPACE real-time control system.

The PID parameters of the servo system are first tuned as $K_p = 0.4$, $K_i = 0.001$, and $K_d = 0.001$, so that the system can respond to the reference quickly. Figure 8.44 shows the control response for a given step, where the dotted line

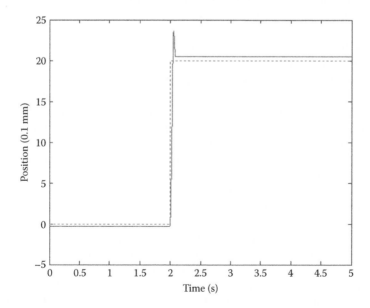

FIGURE 8.44
Step response with PID control.

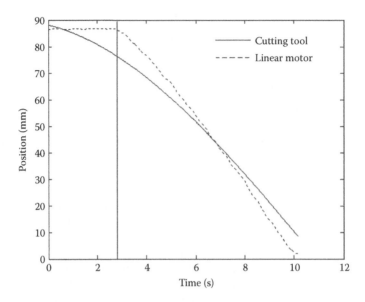

FIGURE 8.45
Controller response and position.

represents the set point, while the solid line represents the actual response. In Figure 8.44, a steady-state error (< 0.1 mm) is observed, due to a small integral gain used for rapid response. For a series of image position information fed to the control system, the system response is shown in Figure 8.45, which compares the position of the machine (green line) and the response of the controller (dotted line). Figure 8.46 shows the tracking error.

The predictive model is used to estimate actual position. The delay time is chosen as 0.2 s, which is enough to include the delay of the image processing. Initially, the linear motor is positioned at 86 mm and then follows the image position x_d at time $T = 2.63$ s. The tracking error is within 10 mm. After the initial phase, all the errors are within 8 mm. Since the target area is a circle with diameter 8.9 mm, the tracking error is acceptable and the sensor can measure the temperature accurately in this range.

The part-simulated experiment setup is shown in Figure 8.47. Data are gathered from two separate experiments. One was done under normal conditions, while the other one was done under abnormal conditions. The environmental temperature was 22°C. Figures 8.48 and 8.49 show the two sets of temperature measurements obtained. It is observed that in experiment 1 the temperature measurement is within 110 to 145°C. In experiment 2, there is an obvious temperature rise at time 40. The peak temperature reached is 170°C, triggering an alarm as the threshold of 150°C is set to detect abnormal situations. The part-simulated experiment proves the efficiency of our

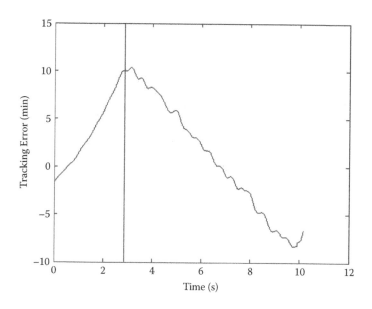

FIGURE 8.46
Controller response and tracking error.

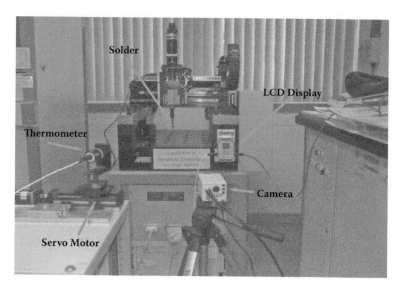

FIGURE 8.47
Simulation scene.

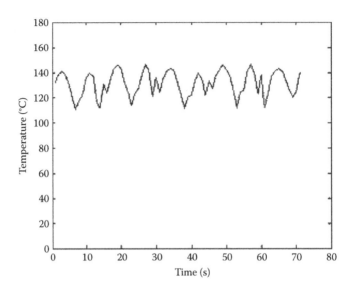

FIGURE 8.48
Temperature measurements under normal conditions.

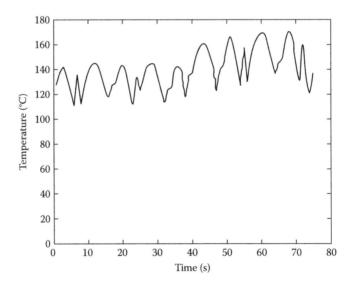

FIGURE 8.49
Temperature measurements under abnormal conditions.

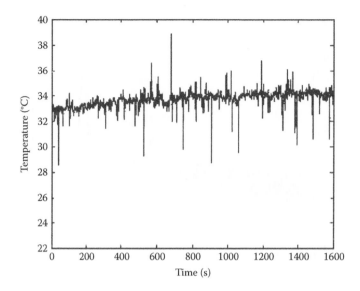

FIGURE 8.50
Temperature measurements for ball-milling machine.

monitoring system in real-time tracking and temperature reading. In the second part, the monitoring system was implemented on a ball-milling and a face-milling machine. The ball-milling machine generates less heat than the face-milling machine. For ball milling, the temperature measurements are plotted in Figure 8.50. The environmental temperature for this experiment is 20°C. In Figure 8.50, the average temperature for a ball-milling machine is around 33°C with a slight temperature rise over time, consistent with extended use of the tool, a phenomenon also observed in Figure 8.51. In Figure 8.51, there is a sudden temperature rise at time 5500 due to the cutter penetrating the workpiece at this time, as explained in Figure 8.52.

In the final stage of the experiments, a thermal camera is used to verify the accuracy of the temperature measurements thus obtained. Results will be shown for the face-milling machine. Four thermal images were recorded during the milling process. Temperature measurements were listed for each tested spot.

In the second stage of the experiments, when a thermometer is used to measure the temperature of the face-milling machine, spot 2 is tracked as in Figure 8.56. Comparing with the results from the thermal camera, the temperature of the spot was kept within 60 to 76°C during milling (the same spot as spot 3 in Figure 8.53, spot 2 in Figure 8.54, and spot 3 in Figure 8.55), which was in accordance with the data furnished in Figure 8.51.

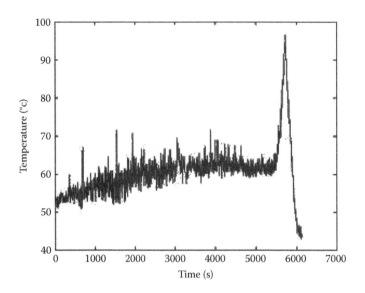

FIGURE 8.51
Temperature measurements for face-milling machine.

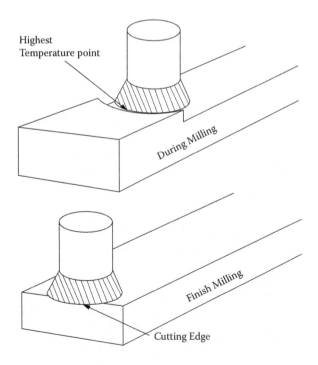

FIGURE 8.52
Explanation of sudden temperature rise.

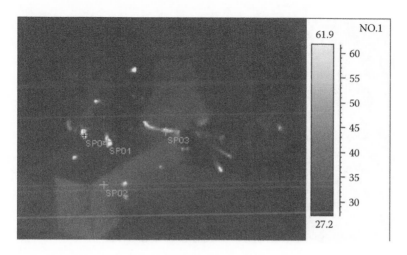

FIGURE 8.53
Accuracy testing using thermal camera (SP01, 115.0°C; SP02, 37.8°C; SP03:69°C; SP04, 127.6°C). Milling processing: phase 1.

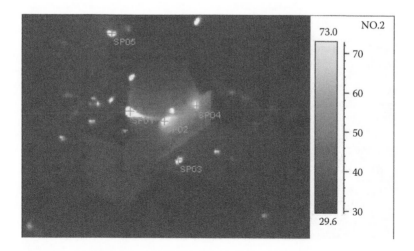

FIGURE 8.54
Accuracy testing using thermal camera (SP01, 169.4°C; SP02, 70.8°C; SP03, 121.0°C; SP04, 58.9°C; SP05, 78.1°C). Milling processing: phase 2.

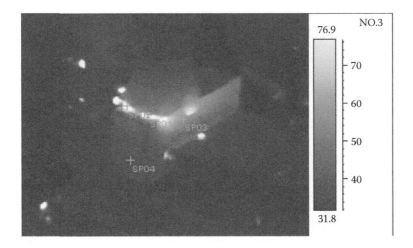

FIGURE 8.55
Accuracy testing using thermal camera (SP01, 115.0°C; SP02, 106.6°C; SP03, 60.3°C; SP04, 39.2°C).
Milling processing: phase 3.

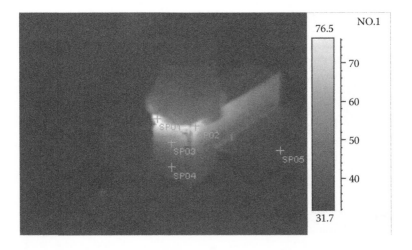

FIGURE 8.56
Accuracy testing using thermal camera (SP01, 87.2°C; SP02, 76.6°C; SP03, 54.4°C; SP04, 41.2°C;
SP05, 35.2°C). Milling processing: phase 4.

References

1. C. V. Newcomb and I. Flinn. Improving the linearity of piezoelectric ceramic actuators. *Electronics Letters*, 18(11), 442–444, 1982.
2. N. W. Hagood, W. H. Chung, and A. von Flotow. Modeling of piezoelectric actuator dynamics for active structure control. *Journal of Intelligent Material Systems and Structures*, 1, 327–354, 1990.
3. D. G. Cole and R. L. Clark. Adaptive compensation of piezoelectric sensoriactuators. *Journal of Intelligent Material Systems and Structures*, 5, 665–672, 1994.
4. J. S. Vipperman and R. L. Clark. Hybrid analog and digital adaptive compensation of piezoelectric sensoriactuators. In *Proceedings of AIAA/ASME Adaptive Structures Forum*, New Orleans, LA, 1995, pp. 2854–2859.
5. P. Ge and M. Jouaneh. Tracking control of a piezoelectric actuator. *IEEE Transactions on Control Systems Technology*, 4, 209–216, 1996.
6. Y. Stepanenko and C. Y. Su. Intelligent control of piezoelectric actuators. In *Proceedings of 37th IEEE Conference on Decision and Control*, Tampa, FL, 1998, pp. 4234–4239.
7. J. M. Cruz-Hernandez and V. Hayward. On the linear compensation of hysteresis. In *Proceedings of 36th IEEE Conference on Decision and Control*, San Diego, CA, 1997, pp. 1956–1957.
8. G. H. Choi, J. H. Oh, and G. S. Choi. Repetitive tracking control of a coarse-fine actuator. In *Proceedings of the 1999 IEEE/ASME International Conference on Advanced Intelligent Mechatronics*, Atlanta, GA, 1999, pp. 335–340.
9. K. Santa, M. Mews, and M. Riedmiller. A neural approach for the control of piezoelectric micromanipulation robots. *Journal of Intelligent and Robotic Systems*, 22, 351–374, 1998.
10. G. Palermo, H. Joris, M. P. Derde, and A. C. Van Steirteghem. Pregnancies after intracytoplasmic injection of single spermatozoon into an oocyte. *Lancet*, 340, 17–18, 1992.
11. K. Yanagida, H. Katayose, H. Yazawa, Y. Kimura, K. Konnai, and A. Sato. The usefulness of a piezo-micromanipulator in intracytoplasmic sperm injection in humans. *Human Reproduction*, 14(2), 448–453, 1998.
12. R. Gilles, D. Dragan, and S. Nava. Separation of nonlinear and friction-like contributions to the piezoelectric hysteresis. *Proceedings of the 2000 12th IEEE International Symposium on Applications of Ferroelectrics*, Honolulu, HI, 2000, pp. 699–702.
13. T. S. Low and W. Guo. Modeling of a three-layer piezoelectric bimorph beam with hysteresis. *Journal of Microelectromechanical Systems*, 4(4), 230–237, 1995.
14. M. Goldfarb and N. Celanovic. Modeling piezoelectric stack actuators for control of micromanipulation. *IEEE Control Systems*, 17, 69–79, 1997.
15. C. Canudas De Wit, H. Olsson, K. Astrom, and P. Lischinsky. A new model for control of systems with friction. *IEEE Transactions on Automatic Control*, 40(3), 419–425, 1995.
16. H. van der Wulp. Piezo-driven stages for nanopositioning with extreme stability: Theoretical aspects and practical design considerations. Delft, The Netherlands: Delft University Press, 1997.

17. Y. Kimura and R. Yanagimachi. Intracytoplasmic sperm injection in the mouse. *Biology of Reproduction*, 52, 709–720, 1995.
18. K. K. Tan, S. C. Ng, and Y. Xie. Optimal intracytoplasmic sperm injection with a piezo micromanipulator. In *Proceedings of the 4th World Congress on Intelligent Control and Automation*, 2002, pp. 1120–1123.
19. J. J. E. Slotine and W. Li. *Applied nonlinear control*. Englewood Cliffs, NJ: Prentice-Hall, 1991.
20. P. A. Ioannou and J. Sun. *Robust adaptive control*. Upper Saddle River, NJ: Prentice-Hall, 1996.
21. R. D'eredit, R. R. Marsh, S. Lora, and K. Kazahaya. A new absorbable pressure-equalizing tube. *Otolaryngology—Head and Neck Surgery*, 127(1), 68–72, 2002.
22. J. Aernouts, J. Soons, and J. J. Dirckx. Quantification of tympanic membrane elasticity parameters from *in situ* point indentation measurements: Validation and preliminary study. *Hearing Research*, 263(1–2), 177–182, 2010.
23. L. C. Kuypers, W. F. Decraemer, and J. J. Dirckx. Thickness distribution of fresh and preserved human eardrums measured with confocal microscopy. *Otology and Neurotology*, 27(2), 256–264, 2006.
24. H. Golnabi and A. Asadpour. Design and application of industrial machine vision system. *International journal of Robotics and Computer-Integrated Manufacturing*, 23, 630–637, 2007.
25. R. Klette and P. Zamperoni. *Handbook of image processing operators*. New York: John Wiley & Sons, 1996.
26. Z. Wang, B.-G. Hu, L. C. Liang, and Q. Ji. Cell detection and tracking for micromanipulation vision system of cell-operation robot. In *Proceedings of IEEE International Conference on Systems, Man, and Cybernetics*, Nashville, TN, October 2000, pp. 1592–1597.
27. T. Bucher et al. Image processing and behavior planning for intelligent vehicles. *IEEE Transactions on Industrial Electronics*, 50(1), 62–75, 2003.
28. D. Karatzasa and A. Antonacopoulos. Colour text segmentation in web images based on human perception. *Image and Vision Computing*, 25(5), 564–577, 2007.
29. C.-S. Cho, B.-M. Chung, and M.-J. Park. Development of real-time vision-based fabric inspection system. *IEEE Transactions on Industrial Electronics*, 52(4), 1073–1079, 2005.
30. P. V. C. Hough. Machine analysis of bubble chamber pictures. In *Proceedings International Conference on High Energy Accelerators and Instrumentation*, 1959, pp. 554–556.
31. A. Kumar. Computer-vision-based fabric defect detection: A survey. *IEEE Transactions on Industrial Electronics*, 55(1), 348–363, 2008.
32. C. L. Herry, M. Frize, and R. A. Goubran. Segmentation and landmark identification in infrared images of the human body. In *Engineering in Medicine and Biology Society 2006 (EMBS '06): 28th Annual International Conference of the IEEE*, August 2006, pp. 957–960.
33. S.-Y. Cho and J.-H. Shim. A new micro biological cell injection system. In *Proceedings of IEEE/RSJ International Conference on Intelligent Robots Systems*, Sendai, Japan, September 28–Oct. 2, 2004, pp. 1642–1647.
34. B.-J. You, Y. S. Oh, and Z. Bien. A vision system for an automatic assembly machine of electronic components. *IEEE Transactions on Industrial Electronics*, 37(5), 349–357, 1990.

35. H. Song and F. Shi. A real-time algorithm for moving objects detection in video images. Presented at *Proceedings of the 5th World Congress on Intelligent Control and Automation*, Hangzhou, People's Republic China, June 15–19, 2004.
36. J. Barron, D. Fleet, and S. Beauchemin. Performance of optical flow techniques. *International Journal of Computer Vision*, 42–77, 1994.
37. P. Rosin and T. Ellis. Image difference threshold strategies and shadow detection. In *Proceedings of British Machine Vision Conference*, 1995, pp. 347–356.
38. Y. Zhang and K. K. Tan. Text extraction from images captured via mobile and digital devices. Presented at *Proceedings of 5th International Conference on Industrial, Automotive*, Montreal, Quebec, Canada, June 11–13, 2007.

Index

Printed and bound by CPI Group (UK) Ltd, Croydon, CR0 4YY

18/10/2024

01776257-0005